Nonlinear Programming
Sequential Unconstrained Minimization Techniques

Classics in Applied Mathematics

Editor: Robert E. O'Malley, Jr.
Rensselaer Polytechnic Institute
Troy, New York

Classics in Applied Mathematics is a series of textbooks and research monographs that were once declared out of print. SIAM is publishing this series as a professional service to foster a better understanding of applied mathematics.

Classics in Applied Mathematics

Lin, C. C. and Segel, L. A., Mathematics Applied to Deterministic Problems in the Natural Sciences

Belinfante, Johan G. F. and Kolman, Bernard, A Survey of Lie Groups and Lie Algebras with Applications and Computational Methods

Ortega, James M., Numerical Analysis: A Second Course

Fiacco, Anthony V. and McCormick, Garth P., Nonlinear Programming: Sequential Unconstrained Minimization Techniques

Nonlinear Programming
Sequential Unconstrained Minimization Techniques

Anthony V. Fiacco
George Washington University

Garth P. McCormick
George Washington University

siam.

Society for Industrial and Applied Mathematics
Philadelphia

Library of Congress Cataloging-in-Publication Data
Fiacco, Anthony V.
  Nonlinear programming : sequential unconstrained minimization
techniques / Anthony V. Fiacco, Garth P. McCormick.
    p. cm. - - (Classics in applied mathematics)
  "This SIAM edition is an unabridged, corrected republication of
the work first published by Research Analysis Corporation, McLean,
Virginia" - - T.p. verso
  Includes bibliographical references (p.).
  ISBN 0-89871-254-8
  1. Nonlinear programming. 2. Mathematical optimization.
3. Algorithms. I. McCormick, Garth P. II. Title. III. Series.
T57.8.F54   1990
519.7'6- -dc20   90-9658
                                                               CIP

Copyright © 1990 by the Society for Industrial and Applied Mathematics

All rights reserved. Printed in the United States of America. No part of this book may be reproduced, stored, or transmitted in any manner without the written permission of the Publisher. For information, write the Society for Industrial and Applied Mathematics, 3600 University City Science Center, Philadelphia, Pennsylvania 19104-2688.

Preparation of this book was undertaken in the course of research sponsored by contract DA-44-188-ARO-1 between the U.S. Army Research Office and the Research Analysis Corporation, Library of Congress Catalog Card Number: 68-30909 SBN 471 25810 5

This SIAM edition is an unabridged, corrected republication of the work first published by Research Analysis Corporation, McLean, Virginia.

To Our Parents

Preface

The primary purpose of this book is to provide a unified body of theory on methods of transforming a constrained minimization problem into a sequence of unconstrained minimizations of an appropriate auxiliary function. The auxiliary functions considered are those that define "interior point" and "exterior point" methods which are characterized respectively according to whether the constraints are strictly satisfied by the minimizing sequence. Initial emphasis is on generality, and the central convergence theorems apply to the determination of local solutions of a nonconvex programming problem. Strong global and dual results follow for convex programming; a particularly important example is the fact that the exterior point methods do not require the Kuhn-Tucker constraint qualification [77] to ensure convergence or characterize optimality.

In addition to giving a rather comprehensive exposition of the rich theoretical foundation uncovered for this class of methods, we wish to emphasize the demonstrated practical applicability of their various realizations. This has been brought about largely by the adaptation and further development of effective computational algorithms for calculating an unconstrained minimum and by the development of special extrapolation techniques for accelerating convergence. Significant progress has also been made in the development of computational techniques that exploit the special structures characterizing large classes of problems. In addition to these efficiencies, which have been effected for the method proper, some exploration has been done in combining the present methods with other mathematical programming algorithms to obtain even more efficient composite algorithms. Various computer programs at the Research Analysis Corporation, some operational since 1961, have been constantly utilized and improved and have provided extensive computational experience.

We have also attempted to provide some historical perspective for the basic approach with an effort toward synthesis. Formal derivations and ample intuitive arguments and simple numerical examples are interjected to motivate and clarify the basic techniques.

Finally, we have attempted to recapitulate the basic supporting developments in the theory of mathematical programming in a direct and simple manner. The satisfaction of the hypotheses of Farkas' lemma [44] is the motivation for various regularity assumptions on the problem functions. This leads directly to deducing the usual necessary conditions for optimality—in particular, the existence of finite Lagrange multipliers. Our presentation is intended to clarify this point of view, which often appears to be clouded in the literature; for example, the Kuhn-Tucker constraint qualification [77] is one such assumption that has often been incorrectly regarded as being essential to characterize optimality. We proceed from those results associated with assuming continuity of the problem functions to those involving first-and second-order differentiability. Several new necessary and sufficient conditions are given for the latter.

This book is the result of several years of extensive research and computational experimentation accomplished initially at The Johns Hopkins Operations Research Office, Bethesda, Maryland, and continued at the Research Analysis Corporation, McLean, Virginia. Initial efforts led to devising a variety of heuristic gradient methods, but they generally proved too slow and unreliable. In 1961 attention was directed to an auxiliary function idea proposed by C. W. Carroll [21]. Our theoretical validation of this approach then led to numerous extensions and generalizations and an efficient computational procedure, which provided the nucleus of the framework for the developments reported here.

Chapters 1, 2, and 7 contain supporting and supplementary material and are included primarily for perspective and completeness. Most of the results of Chapter 2, with the exception of several recently developed second-order necessary and sufficient conditions for isolated and nonisolated constrained minima, are well known. The basic theoretical results for the sequential methods are contained in Chapters 3 and 4. With a few exceptions, the development proceeds in the direction of decreasing generality, from the nonconvex local results of Chapters 3-5 to the convex global results of Chapter 6. Chapter 5 is essentially an in-depth analysis of convergence properties when certain additional conditions hold. Chapter 6 is virtually a recapitulation of Chapters 3-5, the results being recast and strengthened with the important additional assumption of convexity. The computational considerations associated with unconstrained minimization algorithms are relegated entirely to Chapter 8.

In addition to giving many new and more general results, we hope that the book may provide some much-needed clarification and unification in auxiliary function methodology, which only recently has been under significant development. The results for nonconvex problems are among the few that exist in this relatively intractable area and may hopefully be followed by computational breakthroughs as well. Finally, we hope that this exposition will lead to wider recognition of the extremely rich and fertile theoretical basis and the generally proved effective computational applicability of the methods in this class of procedures.

Preface

This book is intended for use by virtually anyone involved with mathematical programming theory or computations as a comprehensive reference for the evolution, theory, and computational implementation of auxiliary function sequential unconstrained methods and a concise reference for mathematical programming theory, several other interesting and effective methods, and some of the most recent advances in the theory and implementation of methods for unconstrained minimization. It could provide a fairly complete basis for an extensive course in mathematical programming or optimization at a wide variety of levels. The nature of the approach generally makes it possible to apply classical methods of analysis, a feature not shared by most techniques and one that can be easily exploited for pedagogical purposes. A complete understanding of the proofs and developments probably requires a solid introduction to analysis and linear algebra. A great portion of the text, however, is motivational and discursive material and requires little more than a brief exposure to elementary algebra. A significant portion depends only on basic notions encountered in first-year calculus and matrix theory.

Many of the results have appeared in previous publications [47-54, 84, 85] as the theory was developed, but a considerable number of significant results appear here for the first time.

Applications of these theoretical developments to the solution of important practical nonlinear programming models are contained in another RAC Research Series book by Jerome Bracken and Garth P. McCormick, "Selected Applications of Nonlinear Programming".

We should like to thank Dr. Nicholas M. Smith for his continued personal interest and effective technical direction. Sustained support for this work has come from the Army Research Office. Professor J.B. Rosen, Dr. James E. Falk, Professor A. Charnes, Dr. Jerome Bracken, Mr. W. Charles Mylander, III, and Professor C.E. Lemke have contributed to the development of this work, particularly Professor Lemke, who reviewed an earlier version of the manuscript and gave numerous helpful suggestions. Special thanks go to Mrs. L. Zazove, who patiently typed several versions of the manuscript.

Anthony V. Fiacco
Garth P. McCormick

McLean, Virginia

Preface to the Classic Edition

The first edition of this book has long been out of print. The primary motivation for its reissue is the current interest in interior point methods for linear programming sparked by the 1984 work of N. Karmarkar. The connection of his projective scaling method with SUMT was pointed out early by Gill and others. The recent work in affine scaling, path following and primal-dual methods bears even more resemblance to our earlier work.

Our initial intent was not to develop a method for linear programming, but for the general nonlinear programming problem. The original book contains some material which was published elsewhere in the open literature, but much of it appears only in the book. In particular, the analysis concerning the trajectory of unconstrained minimizers of the logarithmic barrier function is similar to recent work and appears only here.

Many areas of research were started in this book: e.g. the relationship of penalty function theory to duality theory; use of directions of nonpositive curvature to modify Newton's method when the Hessian matrix is indefinite; integration of the first and second order optimality conditions into convergence and rate of convergence analysis of algorithms; the identification of factorable functions as an important class in applied mathematics; and the beginning of the very important area of sensitivity analysis in nonlinear programming.

Except for corrections, the revision is exactly as originally published. The most important corrections are the proofs of Theorems 6 and 7. Since its original publication the term "penalty function" is now commonly called a "barrier function". This and a few other usages were kept as in the original.

The book was awarded the Lanchester Prize in Operations Research for the year 1968. We hope that readers will find it as fresh and interesting and valuable now as it was then.

Anthony V. Fiacco
Garth P. McCormick
Department of Operations Research
School of Engineering and Applied Science
George Washington University
Washington, D. C.

Symbols and Notations

x	$\equiv (x_1, \ldots, x_n)^T$, an n by 1 vector of variables.
$(E^n)^+$	the nonnegative orthant of Euclidian n-space.
$\|x\|$	$= (\sum_{j=1}^n x_j^2)^{1/2}$, the usual Euclidian norm.
$\nabla_x f(x^k)$,	(sometimes written ∇f^k) is the n by 1 vector whose jth element is $\partial f(x^k)/\partial x_j$.
$\nabla_{xx}^2 f(x^k)$,	(sometimes written $\nabla^2 f^k$) is the n by n matrix whose i,jth element is $\partial^2 f(x^k)/\partial x_i \, \partial x_j$, the Hessian of f at x^k.
Problem A,	minimize $f(x)$ subject to $g_i(x) \geq 0$, $i = 1, \ldots, m$, $h_j(x) = 0$, $j = 1, \ldots, p$.
R	$\equiv \{x \mid g_i(x) \geq 0, \ i = 1, \ldots, m\}$, the region defined by the inequalities of Problem A.
R°	$\equiv \{x \mid g_i(x) > 0, \ i = 1, \ldots, m\}$, the interior of R.
$\mathcal{L}(x, u, w) \equiv$	$f(x) - \sum_{i=1}^m u_i g_i(x) + \sum_{j=1}^p w_j h_j(x)$, the Lagrangian associated with Problem A.
$\alpha(\theta)$,	an arc in E^n parameterized by θ whose tangent at $\theta = \eta$ is denoted by $D\alpha(\eta)$ and whose vector of second derivatives is denoted by $D^2 \alpha(\eta)$.
$A^\#$,	a pseudoinverse of the matrix A that satisfies $AA^\#A = A$.
$\text{diag}(g_i)$,	a diagonal matrix whose ith diagonal element is g_i.
$L(x, r)$	$\equiv f(x) - r \sum_{i=1}^m \ln g_i(x)$, the logarithmic interior point penalty function for Problem B (Problem A with no equality constraints).

$P(x, r)$ $\equiv f(x) + r \sum_{i=1}^{m} 1/g_i(x)$, the inverse interior point penalty function for Problem B (P-function).

$U(x, r)$ $\equiv f(x) + s(r) I(x)$, general interior unconstrained minimization function.

$T(x, t)$ $\equiv f(x) + p(t)O(x)$, general exterior unconstrained minimization function.

$V(x, r, t)$ $\equiv f(x) + s(r)I(x) + p(t)O(x)$, general mixed interior-exterior unconstrained minimization function.

$W(x, r)$ $\equiv f(x) - r \sum_{i=1}^{m} \ln g_i(x) + \sum_{i=m+1}^{q} \dfrac{\{\min[0, g_i(x)]\}^2}{r}$,

the W-function, a mixed interior-exterior unconstrained minimization function for Problem M.

$x^k \to y$ the sequence $\{x^k\}$ converges (strongly, i.e., in norm) to y.

Contents

1 INTRODUCTION
1.1 Statement of the Mathematical Programming Problem   1
1.2 Historical Survey of Sequential Unconstrained Methods for Solving Constrained Minimization Problems   4

2 MATHEMATICAL PROGRAMMING–THEORY
2.1 First-Order Necessary Conditions   17
2.2 Second-Order Necessary Conditions   25
2.3 Second-Order Sufficiency Conditions   30
2.4 Sensitivity Analysis in Nonlinear Programming   34
2.5 Historical Remarks   38

3 INTERIOR POINT UNCONSTRAINED MINIMIZATION TECHNIQUES
3.1 Introduction—Derivation of Algorithms from Sufficiency Conditions for Constrained Minima   39
3.2 General Statement of Interior Point Minimization Algorithms and Their Intuitive Basis   42
3.3 Convergence Proofs for Interior Point Algorithms   45

4 EXTERIOR POINT UNCONSTRAINED MINIMIZATION TECHNIQUES
4.1 General Statement of Exterior Point Algorithms and Their Intuitive Derivation   53
4.2 Convergence Theorem for Exterior Point Algorithms   57
4.3 Mixed Interior Point-Exterior Point Algorithms   59
4.4 Generalized Interior and Exterior Point Method   65
4.5 A Hierarchy of Penalty Functions   68

5 EXTRAPOLATION IN UNCONSTRAINED MINIMIZATION TECHNIQUES
5.1 Trajectory Analysis for Interior Point Techniques   72
5.2 Analysis of Isolated Trajectory   76
5.3 Analysis of Isolated Trajectory of Exterior Point Algorithm   82
5.4 Trajectory Analysis in Mixed Interior Point-Exterior Point Algorithms   84

6 CONVEX PROGRAMMING
- 6.1 Convexity—Definitions and Properties   87
- 6.2 Convex Programming—Theory   89
- 6.3 Solution of Convex Programming Problems by Interior Point Unconstrained Minimization Algorithms   94
- 6.4 Solution of Convex Programming Problem by Exterior Point Unconstrained Minimization Algorithms   102
- 6.5 Additional Results for the Convex Analytic Problem   110

7 OTHER UNCONSTRAINED MINIMIZATION TECHNIQUES
- 7.1 Using Weighted Penalty Functions   113
- 7.2 Q-Function Type Interior Point Unconstrained Minimization Algorithms   121
- 7.3 Continuous Version of Interior Point Techniques   126
- 7.4 Dual Method for Strictly Convex Problems   130
- 7.5 Generalized Lagrange Multiplier Technique for Resource Allocation Problems   137
- 7.6 Solution of Constrained Problem by a Single Unconstrained Minimization   143
- 7.7 Combined Unconstrained-Simplicial Algorithm   145
- 7.8 Gradient Projection Method   149

8 COMPUTATIONAL ASPECTS OF UNCONSTRAINED MINIMIZATION ALGORITHMS
- 8.1 Introduction—Summary of Computational Algorithm   156
- 8.2 Minimizing an Unconstrained Function   157
- 8.3 Minimizing the W Function for Convex Programming Problems with Special Structure   178
- 8.4 Acceleration of Extrapolation   188
- 8.5 Other Algorithmic Requirements   191

REFERENCES   196
INDEX OF THEOREMS, LEMMAS, AND COROLLARIES   203
AUTHOR INDEX   205
SUBJECT INDEX   207

1

Introduction

1.1 STATEMENT OF THE MATHEMATICAL PROGRAMMING PROBLEM

The mathematical programming problem is to determine a vector $x^* = (x_1^*, \ldots, x_n^*)^T$ that solves the problem

$$\text{minimize } f(x) \quad (A)$$

subject to
$$g_i(x) \geq 0, \quad i = 1, \ldots, m, \quad (1.1)$$
$$h_j(x) = 0, \quad j = 1, \ldots, p. \quad (1.2)$$

When the problem functions f, $\{g_i\}$, and $\{h_j\}$ are all linear Problem A is called a *linear programming problem*. If any of the functions is nonlinear the problem is called a *nonlinear programming problem*. There are other terms, such as convex, concave, separable, quadratic, and factorable, which may apply to special cases of Problem A, and these will be defined later. While all the remarks in this book apply in particular to these special cases, we shall at the outset concern ourselves with problems where f, $\{g_i\}$, and $\{h_j\}$ can take on any form of nonlinearity subject only to continuity and differentiability requirements.

The following is a simple example of a nonlinear programming problem.

Example.
$$\text{minimize } f(x) = |x_1 - 2| + |x_2 - 2|$$

subject to
$$g_1(x) = x_1 - x_2^2 \geq 0,$$
$$h_1(x) = x_1^2 + x_2^2 - 1 = 0.$$

The dashed lines in Figure 1 represent isovalue contours of the objective function; that is, points at which the objective function $f(x)$ has constant value. The feasible region is the set of points that satisfy the constraints of

2 Introduction

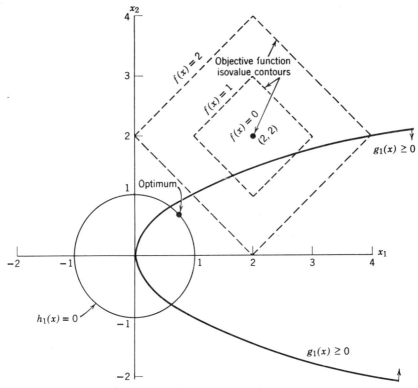

Figure 1 Nonlinear programming example.

the problem. In this example the feasible region is the arc of the circle lying within the parabola. A solution to the problem is any point in the feasible region with smallest objective function isocontour value. This is seen by inspection to be $(\sqrt{2}/2, \sqrt{2}/2)$. If the equality constraint is removed the solution is seen to be at $(2, \sqrt{2})$. If both constraints are removed the solution is at $(2, 2)$. In the latter case $(2, 2)$ is called an *unconstrained* minimum.

Usually the functions of Problem A are required to be continuous. Much of the theory of nonlinear programming concerns the case when the functions are continuously differentiable, or twice continuously differentiable. In these instances it is possible to prove theorems which characterize solutions to Problem A. These theorems in turn influence the development of algorithms for solving the programming problems.

Several classes of mathematical programming problems have been dealt with in recent years. We briefly mention some of these and give a number of standard references that develop theoretical results and computational methods for solving the corresponding problem. The linear programming problem has been treated extensively, and many significant results have been forthcoming, such as effective methods for problems having a particular structure. There exists an enormous literature on the subject. For basic results refer to the list of annotated references given in [96] and to the developments and additional bibliography given in [25, 26, 34, 62].

Particular results and algorithms have been obtained for "quadratic programming," where $f(x)$ is a positive semidefinite quadratic form, and the constraints are linear [7, 8, 110]. Special methods have been developed when $f(x)$ is a convex separable function and the constraints are linear [27, 88]. Special-purpose algorithms also exist for the case where $f(x)$ is convex and the constraints are linear [45, 98].

The case where $f(x)$ is convex, the $g_i(x)$ concave, and the $h_j(x)$ linear has received particular attention. When these conditions prevail (A) is called a convex programming problem. The smoothness of the problem functions makes the problem well behaved, and the convexity-concavity assumptions assure that the feasible region (set of points satisfying the constraints) is convex and, most importantly, that any local solution is also global. The basic optimality conditions for the problem were given in [77]. Numerous other contributions to the theory and development of computational algorithms have appeared, largely in the last decade [1, 6, 28, 39, 49, 50, 65, 70, 74, 99, 111, 118].

The developments mentioned above are often applicable in a "local" sense; that is, they hold if x is restricted to a suitable domain such that the requisite conditions hold in that domain. This means that some results are easily extended to apply to the characterization of *local* solutions of (A), when (A) is a nonconvex problem. The "general" nonconvex problem, where (A) is not even necessarily "locally convex" in any neighborhood of a relative minimum, has remained rather intractable. Most of the results in this important problem area have been theoretical and are quite recent [16, 17, 101, 104, 117].

The group of algorithms called sequential unconstrained minimization techniques has given rise to numerous theoretical results and effective computational procedures for solving the convex programming problem. Recent developments indicate that these results can be generalized and extended significantly, since the basic technique can be validated under very general and weak conditions. Thus a number of important results have been obtained for nonconvex programming, as well as additional generality and a finer characterization for problems having a special structure. A history of sequential unconstrained methodology is contained in the next section.

This book pursues the development of these sequential unconstrained methods.

1.2 HISTORICAL SURVEY OF SEQUENTIAL UNCONSTRAINED METHODS FOR SOLVING CONSTRAINED MINIMIZATION PROBLEMS

The Transformation Approach

The methods we shall discuss are based on transforming a given constrained minimization problem into a sequence of unconstrained minimization problems. This transformation is accomplished by defining an appropriate auxiliary function, in terms of the problem functions, to define a new objective function whose minima are unconstrained in some domain of interest. By gradually removing the effect of the constraints in the auxiliary function by controlled changes in the value of a parameter, a sequence or family of unconstrained problems is generated that have solutions converging to a solution of the original constrained problem.

For simplicity in the present discussion, we proceed formally to sketch the basic idea. The problem is to find a solution x^* of

$$\text{minimize } f(x) \tag{B}$$

subject to

$$g_i(x) \geq 0, \quad i = 1, 2, \ldots, m,$$

where $x \in E^n$

A typical unconstrained auxiliary function may have the form

$$\varphi[x, \lambda(t)] \equiv f(x) + \sum_{i=1}^{m} \lambda_i(t) G[g_i(x)],$$

where t is a parameter, the $\{\lambda_i(t)\}$ are weighting factors, and $G(y)$ is generally a monotonic function of y that behaves in some well-chosen manner at $y = 0$. Typical choices are either that $G(y) > 0$ for $y < 0$ and $G(y) = 0$ for $y \geq 0$, or that $G(y) \to +\infty$ as $y \to 0^+$. The former choice usually is associated with procedures that are not concerned with constraint satisfaction except at the solution, and the latter, where constraint satisfaction is enforced throughout.

When successful, the method generally proceeds computationally as follows. Select a sequence $\{t_k\}$ such that $t_k \geq 0$, and $t_k \to \infty$ as $k \to \infty$. Compute a minimum x^k of $\varphi[x, \lambda(t_k)]$ for $k = 1, 2, \ldots$ Under appropriate conditions such an x^k exists and is an unconstrained minimum of $\varphi[x, \lambda(t_k)]$. Usually the most desirable result is that $\lim_{k \to \infty} x^k = x^*$, a solution of (B). A weaker result, often adequate, is that $f(x^k) \to f(x^*)$, a minimum value of the objective function. Invariably, the result follows that

$$\lim_{k \to \infty} \sum_{i=1}^{m} \lambda_i(t_k) G[g_i(x^k)] = 0,$$

1.2 Historical Survey of Sequential Unconstrained Methods

so that also
$$\lim_{k \to \infty} \varphi[x^k, \lambda(t_k)] - f(x^*) = 0;$$

that is, the modified objective function converges to the same minimal value as the original objective function. This means, in effect, that the influence of the constraints on the modified objective or auxiliary function is gradually relinquished and finally removed in the limit.

This, then, is the general idea of the approach. Its great advantage lies in the fact that the constraints need not be dealt with separately and that classical theory and modern methods for computing unconstrained extrema can be brought to bear. The theoretical difficulties are in such areas as prescribing general conditions that guarantee convergence of either the minimizing sequence or the associated modified objective function values, and in validating acceleration procedures. Computationally, obtaining rapid convergence is a central concern and depends on the efficacy of methods for unconstrained minimization and on procedures for effective extrapolation. In recent years a substantial body of theory has been established and effective computational algorithms have been implemented [49, 51].

We turn to a chronological account of the origin and development of the technique of solving a constrained problem by transforming it into a sequence of unconstrained problems. Before pursuing this relatively recent history, a note on the classical Lagrange multiplier procedure is particularly in order.

The Lagrange Multiplier Technique

Our primary interest is in sequential unconstrained methods for solving a problem of type (B). However, it should be remarked at the outset that the Lagrange multiplier technique (see [108], for example) for handling problems of type (E) [minimize $f(x)$ subject to $h_j(x) = 0$, $j = 1, \ldots, m$] is surely a technique for transforming the problem into an unconstrained problem. Note that this simply amounts to the choices $\lambda_j(t) = \lambda_j$ (constant) and $G(y) = y$ in $\varphi[x, \lambda(t)]$. In fact, by introducing appropriate slack variables, we can transform any problem of type (B) into a problem of type (E) and then formulate the associated Lagrangian problem. This procedure has been applied to variational problems [11, 63, 106].

We do not propose to delve into a detailed description of the classical Lagrange multiplier technique, or into a discussion of the relative merits of the procedure or of the computational difficulties inherent in a direct application of it to a particular example. These matters have been treated in some detail elsewhere [23, 37, 76]. We simply wish to indicate here that the Lagrangian is a classical example of the unconstrained auxiliary function approach.

6 Introduction

Recent developments on the Lagrange multiplier technique are discussed in Sections 7.4 and 7.5. This is based on the work in [40] and [43].

Besides being viewed directly as a special case of the type of auxiliary function we are considering, it follows that the Lagrangian is inextricably associated with every method for mathematical programming, because conditions for characterizing solutions of mathematical programming problems such as (B) or (E) directly involve the associated Lagrangian. This may be viewed as a direct consequence of the fact that, assuming differentiability, a necessary condition that x^* solve (E) is that x^* be a stationary point of the associated Lagrangian for some real λ_i, constants to be determined from the requirement of stationarity and the relations $h_j(x) = 0$, $j = 1, \ldots, m$. Thus the Lagrangian will always be very much involved in our theoretical results, as much of the text will amply testify.

Owing to the additional fact that the Lagrangian may be viewed as a special case of the auxiliary unconstrained function, $\varphi[x, \lambda(t)]$, it is not surprising that we shall see extremely close connections throughout between the associated Lagrangian of a mathematical programming problem and the particular auxiliary function being utilized to transform that problem into a sequence of unconstrained minimizations.

The following historical account is intended to be indicative of the main stream of developments. Inclusions and omissions reflect the authors' point of view and, likewise, interpretations and appraisals are subjective.

Chronological Survey of Developments

In 1943 R. Courant [29] suggested studying the conditions for stationarity of $f(x) + tg^2(x)$ as $t \to \infty$, to analyze motion constrained to satisfy $g(x) = 0$ in terms of unconstrained motion. The suggestion was motivated by physical considerations and was not offered directly as a technique for solving a mathematical programming problem. The idea was apparently not rigorously pursued for over a decade.

In the interim several important theoretical developments took place. In 1951 Dantzig [33] formalized the linear programming problem and offered a first version of the simplex method. An enormous amount of effort was subsequently directed toward the development and implementation of linear programming algorithms. Shortly thereafter, in 1951, Kuhn and Tucker [77] published their results on necessary and sufficient conditions characterizing the solution of the nonlinear convex programming problem and gave an equivalence between this problem and the saddle-point problem of the Lagrangian. In 1950 Arrow and Hurwicz [5] treated this latter equivalence as well.

Arrow [4] in 1951 devised a gradient technique for approximating saddle

1.2 Historical Survey of Sequential Unconstrained Methods 7

points and constrained maxima. This is one of the earlier gradient techniques offered for solving the constrained nonlinear problem. It qualifies for a method based on solving an unconstrained problem, in this case, in attempting to satisfy the Lagrangian necessary conditions directly.

In considering various techniques for satisfying a system of linear inequalities of the form $y_i = g_i(x) \geq 0$, $i = 1, \ldots, m$, with $x \in E^n$, T. S. Motzkin [89] in 1952 suggested that various gradient methods could be applied to minimizing $F(y) \equiv \sum_{i=1}^{m} e^{-\lambda_i y_i}$, where $\lambda_i \geq 0$. He pointed out that this should lead to $y_i \geq 0$, all i, and hence to a point x satisfying the original inequalities. Though this particular function was not applied to develop a usable algorithm (to our knowledge), it is an interesting example of how a problem involving several constraints can be converted to an unconstrained problem. Note also that the constraints need not be linear to make this approach applicable.

In 1954 and 1955 there was renewed the serious consideration of the penalty function approach as a computational device. K. R. Frisch [59, 60] introduced his "logarithmic potential method," a method based on utilizing the gradient of the function $f(x) + \sum_{i=1}^{m} \alpha_i \ln g_i(x)$ in the interior of the feasible region, to hasten convergence to the solution of Problem B, which we repeat here for convenience.

$$\text{minimize } f(x)$$
subject to
$$g_i(x) \geq 0, \quad i = 1, \ldots, m.$$

The α_i are specified constants. It is evident that the logarithmic penalty function is one of the class described above. However, Frisch apparently did not minimize this function sequentially, but merely used its gradient, in combination with the objective function gradient, to enforce feasibility and to attempt to accelerate convergence as part of an auxiliary device in a gradient scheme. In particular, the problem of completing convergence to the optimum was not solved. It should be noted that, used sequentially as indicated, the logarithmic function gives rise to an excellent realization of the unconstrained techniques described in this text.

Also in 1955 we see an interesting application of penalty function gradients in solving programming problems on the analog computer. Ablow and Brigham [2] essentially utilized the gradients of $f(x) - t \sum_{i=1}^{m} \min(g_i(x), 0)$, and $f(x) + t \sum_{i=1}^{m} [\min(g_i(x), 0)]^2$ to devise two continuous gradient methods for Problem B [and hence also for (E), since any equality can be written as two inequalities]. Apparently the technique is effective for small nonlinear problems, the solution being approximated when the indicated functions have vanishing gradients for large t. It is interesting that the gradients of the two functions indicated are used, these essentially in a heuristic algorithm. The functions themselves are not explicitly stated, and no convergence proofs are given in the reference cited. It should be noted that both penalty functions give rise to an unconstrained formulation approach, as previously described. In fact, the second function mentioned is the natural extension of Courant's quadratic penalty function idea to multiple inequality constraints. We shall see that this approach, utilized on a digital computer with a discrete sequence $\{t_k\}$ such that $t_k \geq 0$ and $t_k \to +\infty$, gives rise to powerful algorithms, both in variational problems and in finite dimensional problems. A completely rigorous treatment of the idea, for multiple variables and constraints, was not developed until after 1955.

Apparently the second function mentioned in the preceding paragraph was independently utilized and proposed by a number of people at different times. For example, in 1955 G. D. Camp [18] suggested introducing new variables y_i, such that $g_i(x) = y_i^2$, and then applying the quadratic penalty approach to solve (B); that is, sequentially minimize $f(x) + t_k \sum_{i=1}^m [g_i(x) - y_i^2]^2$ for a positive increasing sequence $\{t_k\}$. The conjecture was that the corresponding sequence of minima $\{x(t_k)\}$ would be such that $x(t_k) \to x^*$, a solution of Problem B as $t_k \to +\infty$. It may be observed that minimizing the above function is equivalent to minimizing the second function of the preceding paragraph. We shall see that the proposed technique has considerable merit, but that it was not exploited for several years. The scheme was later shown to lead to successful methods under mild regularity conditions and in general spaces. It gave rise to significant theoretical developments and proved to be computationally effective in both variational and ordinary optimization problems.

Moser [30] in 1956 proved convergence of the technique suggested by Courant. To solve the problem minimize $f(x, y)$ subject to the single constraint $g(x, y) = 0$, the function $f(x, y) + tg^2(x, y)$ is minimized for a sequence of values of t, $\{t_k\}$, such that $t_k \geq 0$ and $t_k \to +\infty$ as $k \to \infty$. The limit points of the respective minima $\{[x(t_k), y(t_k)]\}$ are shown to yield the desired solution under mild and general conditions. The proof is confined to two variables and a single equality constraint, but the underlying space is quite general, so that the technique applies, in particular, to variational problems. Courant and Moser also give examples of the use of such approximation theorems to prove theorems in the calculus of variations. An interesting example of this is the proof of the existence of the Euler-Lagrange multiplier λ such that the usual Lagrangian equation, $\nabla f = \lambda \nabla g$, is satisfied at the solution.

Rubin and Ungar [101] in 1957 finally pursued this approach in greater depth. They studied the stationary character of a functional of the form

$$\int_0^\alpha \left[f(x, \dot{x}) + \mu_k \sum_{i=1}^m g_i^2(x) \right] dt,$$

1.2 Historical Survey of Sequential Unconstrained Methods

where $x = [x_1(t), \ldots, x_n(t)]$, and obtained results in a class of variational problems, thereby showing that motion constrained to satisfy the $g_i(x) = 0$ in a conservative force field can be approximated as the limit of a sequence of unconstrained motions (as $\mu_k \to +\infty$). This is a generalization and validation of the idea for multiple variables and multiple equality constraints. They proved, under suitable regularity conditions, such as the assumption that the constraint gradients are linearly independent, that a sequence of unconstrained motions provides the stationary solutions to a respective sequence of unconstrained functionals, as indicated above. They provided a convergence proof, a differentiability characterization of the solution, and a proof of the existence of Lagrange multipliers such that their solution satisfies the Lagrange equations of the first kind. These results were significant and took the original Courant suggestion out of the realm of conjecture for a much wider class of problems; that is, to problems involving multiple variables and multiple constraints. It appears, however, that the results were not vigorously pursued or applied, perhaps because they were given in the context of an analysis of constrained motion and were not offered with direct emphasis on the problem of optimization.

A very complete and rigorous analysis of the Lagrangian saddle-point approach to solve constrained problems, both linear and nonlinear, was given by Arrow, Hurwicz, and Uzawa [6] in 1958. A number of interesting results were obtained that extended and supplemented the theory of mathematical programming. Computationally, the "differential gradient" schemes considered for generating a sequence of points converging to the required saddle point converged too slowly for practical feasibility. We share the opinion of Wolfe [113], who suggested as recently as 1965 that these approaches may still provide efficient computational procedures if large-step methods can be devised, rather than those techniques which have been tried. The latter have been concerned with closely simulating the precise continuous trajectory that converges to the solution. In many applications the trajectory has little or no value to the user.

In 1959 we encounter again the idea of devising a penalty function that enforces feasibility, a function very closely related to Frisch's logarithmic function mentioned above. This is the function $f(x) + t \sum_{i=1}^{m} [g_i(x)]^{-1}$, proposed by Carroll [20] and utilized by him to obtain an approximate solution of a number of moderately sized problems. In solving Problem B one minimizes this function in the region where $g_i(x) > 0$, all i, for a sequence $\{t_k\}$ of values of the parameter t, such that $t_k > 0$ and $t_k \to 0$. Thus one obtains a corresponding minimum $x(t_k)$, and limit points of $\{x(t_k)\}$ solve B, when the procedure is valid. The usual consequence, that is, the vanishing of the penalty term in the limit, follows necessarily. The proposal was intuitive. No proofs were given, though plausible intuitive arguments were offered, supported by some indication of successful computational results. Theoretical validation and demonstration of computational efficacy was to be furnished after several years, by Fiacco and McCormick in 1963, as we indicate in the following.

In 1960 we encounter an interesting technique suggested by Rosenbrock [100] for automatically assuring that the objective function be reduced, while ensuring feasibility, in a problem of type (B) with nonempty interior. The type of problem considered in [100] is of a very special sort, but the basic idea proposed can be generalized as follows. Suppose we require a feasible interior point x^0 such that $f(x^0) < 0$, assuming that such exists. The idea is based on defining a modified objective of the form

$$f(x)\Psi_1[g_1(x)] \cdots \Psi_m[g_m(x)],$$

where $\Psi_i \geq 0$ are chosen so that the modified function vanishes outside the feasible region and is negative in the interior of the feasible region. The modified function, under reasonable conditions, will have an unconstrained feasible interior minimum at some point x^1, which means $g_i(x^1) > 0$, all i, implying that $f(x^1) < 0$, so that we can take x^1 as our required x^0. It is clear that this approach is intimately related to the approaches based on an auxiliary function of the type that involves a "penalty" term added to $f(x)$. One need only take logarithms of the function given above to see this connection. It is also evident that any technique that can reduce the objective function below a stipulated value and maintain feasibility is a candidate for an optimization procedure. Thus Rosenbrock adapted his technique to such, offering his own modified gradient procedure and evidence of practical implementation [100]. It is of further interest to note what now appears to be a recurring phenomenon: that, considerably later, the sort of technique proposed was made formal, rigorous, and general. In this case the time lapse was about four years, when Huard [73] published his work on "the method of centers."

In 1961 Carroll [21] published a summary report on his inverse penalty function technique, which we have already described above. It was this publication that brought the method to our attention and subsequently stimulated us to investigate it further, both theoretically and computationally [49–51].

Frisch's logarithmic function (see above) was further utilized and developed by Parisot [91] in 1961 to devise sequential unconstrained methods for solving linear programming problems.

Recalling again the basic idea originally suggested by Courant [29] in 1943, aside from the results of Moser [30] in 1950 and Rubin and Ungar [101], obtained in 1957 and mentioned above, we see that there was apparently no extensive significant theoretical development of the approach for some

1.2 Historical Survey of Sequential Unconstrained Methods   11

time. Nonetheless, the quadratic penalty function technique was often used as a computational device to approximate solutions of variational problems, in particular. This development has not been traced out in detail by the authors, but the evidence of it is verified by such articles as appeared in 1962 in the book on optimization techniques edited by Leitmann [78]. In Chapter 1, T. N. Edelbaum refers to "penalty functions," of which the Courant function is the prime example. He suggests such "obvious modifications" as $f(x) + g_1^{2t}(x)$, obtaining the minimizing $x(t)$ and taking $t \to \infty$ as usual, for the solution of a problem of type (E). In Chapter 6 of the same book H. J. Kelly deals rather extensively with the technique and indicates the form of the quadratic penalty function applicable to Problem B. The function given is $f(x) + \sum_{i=1}^{m} K_i g_i^2(x) H(g_i)$, where $H(g_i) = 0$ if $g_i(x) \geq 0$ (that is, for x feasible) and $H(g_i) = 1$ if $g_i(x) < 0$ (x not feasible), and the K_i are positive numbers. For K_i specified, the function is minimized to yield the point $x(K_1, \ldots, K_m)$. As with the equality constraints, the inequality constraints are enforced in the limit as the $K_i \to \infty$. Kelly applied this penalty function approach to approximate the solution of a control problem with state variable constraints.

As a brief aside, we note that the modification indicated above of the quadratic penalty technique applied to inequality constraints is a rather natural application of Courant's function as extended to handle multiple equalities. Since $g_i(x) \geq 0$ is equivalent to $g_i(x) = y_i^2$, the equality version of the penalty function approach requires minimization of

$$f(x) + t \sum_{i=1}^{m} [g_i(x) - y_i^2]^2,$$

as we have already noted. This is equivalent to minimizing

$$f(x) + t \sum_{i=1}^{m} g_i^2(x) H(g_i),$$

with $H(g_i)$ as above, as can be readily seen. The introduction of weights in the latter terms is an obvious modification.

In 1962 we finally see a detailed and rigorous treatment of the Courant idea, as extended to multiple equalities and variables and viewed directly in application to the mathematical programming problem

$$\text{minimize } f(x) \qquad \text{(E)}$$

subject to

$$h_j(x) = 0, \quad j = 1, \ldots, p.$$

This was done by T. Butler and A. V. Martin [17], who were motivated by work such as that of H. J. Kelly [78] on optimization of aircraft flight paths, which they reference in their article. Butler and Martin were concerned with the rigorous development of conditions that assure existence of the desired minimizing sequence and of the limit to the solution point and value, and they also developed interesting and important results that pertain to the consequences of slightly relaxing the constraints. Their main result is that, if S is a topological space, $h_1^2(x), \ldots, h_p^2(x)$ and $f(x)$ are lower semicontinuous real-valued functions defined in S, and $x_n \in S$ minimizes $f_n(x) \equiv f(x) + \sum_{j=1}^{p} K_{n_j} h_j^2(x)$, with K_{n_j} positive constants such that $K_{(n+1)_j} > K_{n_j}$ and $\lim_{n \to \infty} K_{n_j} = +\infty$, then if x^* is any limit point of $\{x_n\}$, $h_j(x^*) = 0$, all j, and x^* solves Problem E (with min possibly replaced by inf in that problem). Thus the desired general approximation theorem for the problem of type (E) is established.

T. Pietrzykowski [92] in 1962 proved the validity of the quadratic penalty function approach (see above) as applied to the convex programming problem [$f(x)$ convex, all $g_i(x)$ concave] in E^n. He showed that any limit point of the sequence of minimizing points is a solution of (B), assuming compactness of the constraint region. There is no tie-in indicated with duality theory, but this was later given by Fiacco and McCormick [54], along with a relaxation of the compactness assumption.

In 1963 we [49] proved the convergence of Carroll's reciprocal penalty function approach (see above) under conditions usual for the convex programming problem in E^n. We showed further that the penalty function minimization furnishes dual feasible points as a by-product. This appears not to have been observed previously for the penalty function approach. Considerable computational experiments were performed, and the practical feasibility of a computational algorithm was demonstrated.

As an indication of interesting possible variations in the basic theme, Goldstein and Kripke [67] showed the convergence of a subsequence of $\{x(t_k)\}$, the minima of $f(x) + (1/k^2) \sum_{i=1}^{k} g_i^{2k}(x)$ for $k = 1, 2, 3, \ldots$, to the solution of minimize $f(x)$ subject to $x \in S \equiv \bigcap_{k=1}^{\infty} S_k$, where

$$S_k \equiv \{x \mid -1 - 1/k \leq g_i(x) \leq 1 + 1/k, i = 1, 2, \ldots, k\}$$

(that is, $S \equiv \{x \mid -1 \leq g_i(x) \leq 1, \text{all } i\}$) and S is assumed to be compact. This could give rise to algorithms based on estimating the problem functions by a family of supporting hyperplanes, as Goldstein and Kripke indicate in [67].

An important variation of the penalty function approach, one that again revolves about a suitably defined auxiliary function whose unconstrained minima converge to the solution of the constrained problem, is that termed the "method of centers" by Huard [73]. We have already sketched the basic idea above in discussing the work of Rosenbrock [100]. The idea can be conveyed perhaps most easily by thinking of the objective function simply as an additional constraint, $f(x) \leq f(x^0)$, where x^0 is feasible, and then devising

1.2 Historical Survey of Sequential Unconstrained Methods

a function that is minimized by a point in the (assumed nonempty) interior of the region $R_0 \equiv \{x \mid g_i(x) \geq 0, \text{ all } i \; f(x) \leq f(x^0)\}$. If successful, we then have $x^1 \in R_0$ such that $f(x^1) < f(x^0)$. We then replace x^0 by x^1 above and repeat the procedure. Ideally we approximate the solution after a number of iterations of this sort. Huard [73] developed the theory of the basic approach and, with Bui Trong Lieu [16], extended its applicability to general spaces.

In 1965 T. Pomentale [93] studied the penalty function $f(x) + t\Psi(x)$ as applied to Problem B, where $\Psi(x) > 0$ and is continuous in the interior of the feasible region, becoming infinite on the boundary. For the convex problem with compact constraint region, he showed that the penalty function value evaluated at a minimizing point $x(t)$, $t > 0$, converges to an optimal solution value of (B) as $t \to 0$. In particular, then, Carroll's inverse penalty function [21] is a special case of this general one.

Fiacco and McCormick [53] in 1966 showed the applicability of the sequential penalty function approach to certain classes of convex programming problems having a mixture of both inequality and equality constraints. The auxiliary function utilized absorbs the inequality constraints via Carroll's inverse penalty function, and the equality constraints by the addition of the Courant-type quadratic penalty function. Convergence to the optimal solution vector is shown, along with duality results and the validation of convergence acceleration by extrapolation.

An interesting realization of a "method of centers" was shown to be valid by us [52] in 1965 for the convex programming problem of type (B) in E^n. The interior of the feasible region must be assumed nonempty for such methods. The particular function utilized is $[f(x^k) - f(x)]^{-1} + \sum_{i=1}^{m} [g_i(x)]^{-1}$, where x^1 is an interior point and it is assumed that the interior of the set $\{x \mid g_i(x) \geq 0, i = 1, \ldots, m, f(x) \leq f(x^1)\}$ is not empty. The function is convex in the interior of this set and, since it is infinite on the boundary of the set, attains a minimum at an interior point. It follows easily that a sequence $\{f(x^k)\}$, $k = 1, 2, 3, \ldots$, is generated such that $f(x^k)$ decreases monotonically to its optimum value, under the usual conditions for convex programming. In fact, the stronger result $x^{k_j} \to x^*$, a solution of (B), holds for any convergent subsequence of $\{x^k\}$. A most interesting additional result is that the minima of this function are also minima of the Carroll inverse penalty function described earlier, for certain choices of the parameter t. This particular function was developed and utilized without the benefit of Huard's theory on methods of centers [73] mentioned previously, although, as indicated, it is a special case of such a method. (This function is indicated as an example of method of centers function in the general results obtained in Chapter 7.)

In 1965 W. I. Zangwill [115, 117] studied the general penalty function having the form $f(x) + tG[g(x)]$ in the space E^n, where $g(x)$ is a q vector whose components, as functions of x, are either zero or non-negative; that is, the problem of concern is

$$\text{minimize } f(x) \tag{F}$$

subject to

$$g_i(x) \geq 0 \quad \text{for} \quad i = 1, \ldots, m,$$

$$g_j(x) = 0 \quad \text{for} \quad j = m+1, \ldots, q.$$

The loss function is assumed continuous; $G(y) = 0$ if y is feasible and is positive otherwise. The other significant assumptions are the continuity of the problem functions, the assumption that the auxiliary function attains a minimum for t large enough, and the assumption that the feasible region is properly contained in a bounded set. The algorithm proceeds, as usual, minimizing the auxiliary function over $\{t_k\}$ such that $t_k \geq 0$ and $t_k \to \infty$, to generate a corresponding minimizing sequence $\{x^k\}$. The main result is the existence of a subsequence converging to x^*, a solution of (F). Zangwill obtained a dual relationship without convexity assumptions, and other results with convexity and differentiability conditions. Note that the Courant quadratic penalty function is a special case of Zangwill's function. A very intriguing result is the fact that the convex programming problem of form (F) is shown to be solved by the unconstrained minimum of a particular auxiliary function when the parameter t is large enough. (This is contained in Section 7.6.) Zangwill's work provides a very general treatment of the quadratic penalty function approach to constrained problems in E^n.

In 1965 R. E. Stong [104] proved the validity of the Fiacco-McCormick procedure based on Carroll's inverse penalty function (see above) for general topological spaces. The central assumptions are continuity of the problem functions and compactness of the feasible region, from which the convergence of the values of the penalty function, corresponding to any *global* minimizing feasible interior sequence of points, to the *optimal* value of Problem B (with nonempty interior) is deduced. With the additional assumption that the feasible region is the closure of its interior, it follows that every sequence of global minimizing points of the auxiliary function contains a subsequence converging to a global optimum point. This constitutes a significant generalization of our results [50] for the Carroll procedure [21] and is analogous to the development effected by Butler and Martin [17] for the Courant quadratic penalty function approach.

In 1965 also we see evidence of the significant application of the quadratic penalty function approach, generalized for multiple variables and multiple inequality constraints. We refer to the work of Beltrami and McGill [9], who applied the method to constrained variational problems in the theory of search. They employed an interesting generalization of the Newton-Raphson

1.2 Historical Survey of Sequential Unconstrained Methods

procedure to operator equations in more general spaces (see McGill and Kenneth [87] for a description of this and for additional references) to obtain what they termed an "effective computational method." Interesting use was also made of Pontryagin's maximum principle [94]; that is, in admitting candidates for optimality rather than in testing for optimality. It is thus quite clear that the penalty function approach has proved to be effective for variational problems of the nonclassical variety.

In 1967 Fiacco and McCormick [54] further developed the quadratic penalty function approach, applying it to the convex programming problem in E^n, where both equality and inequality constraints are permitted. The essential assumption, other than the convexity condition, is the assumption that the set of optimum points of the problem is compact. The procedure was shown to generate a sequence of minimizing points such that every limit point is a solution of the problem. Duality results were obtained. An important fact is that the interior of the feasible region may be empty, implying that the Kuhn-Tucker constraint qualification [77]—a regularity condition invariably invoked—is not required in order to prove convergence and, hence, optimality. This work was an extension of the results of F. Pietrzykowski [92] mentioned above. Significant computational results by this method have only recently been obtained, but it is clear that the technique will prove computationally very effective for nonlinear programming.

Recent and Present Developments

The penalty functions indicated above are special cases of the general penalty functions developed in Chapters 3 and 4 of this book. Convergence proofs are established for obtaining local solutions of the nonconvex problem in E^n under conditions that are generally weaker than those previously assumed. A theoretical basis for an extrapolation procedure for accelerating convergence is given.

Further generalizations and extensions of a number of the above results have recently been obtained by Fiacco [48]. The theoretical basis for new algorithms is presented, based on modifications of the more general penalty functions that are defined. The method of centers (see above) and the sequential penalty function methods are shown to be directly related and, in fact, large classes of these are shown to give rise to equivalent procedures. This equivalence leads to new methods, including a new class of methods of centers. Assumptions of local analyticity and convexity provide a stronger characterization of the general penalty functions for a wide class of problems.

Finally, for the strictly convex problem, Falk [42] has given the theoretical validation of a sequential method of the general class described, based essentially on defining a penalty function in terms of a method of centers function. Feasibility is not enforced, and it develops that the terminal choice of the involved parameter is not known in advance. The method provides an interesting variation on a "primal-dual" or Lagrangian approach [41]: a sequence of Lagrange multipliers converging to an optimal multiplier and, simultaneously, a sequence of points converging to a solution point are determined by performing a sequence of iterations based on first selecting a parameter value in the auxiliary function, then determining a corresponding unconstrained minimum of the resulting function, using the minimum to update the parameter value, and continuing in this manner. A summary of this material is contained in Section 7.4.

2

Mathematical Programming—Theory

Although one may be at times interested only in optimum or "global" solutions to mathematical programming problems, in general one can only prove theorems about *local* solutions. A local solution is a point at which, in a neighborhood about that point, there is no other point satisfying the constraints that gives a smaller value of the objective function. A global solution can be defined as any local solution yielding the smallest objective function value. If the problem functions have special properties, for example, if they define a convex programming problem, it is possible to prove that every local solution is a global solution (see Chapter 6). Useful theorems stipulating a set of necessary conditions and a set of sufficient conditions that a point be a local solution to Problem A can be developed without recourse to these special properties. The remainder of this chapter is devoted to development of these sets of conditions.

2.1 FIRST-ORDER NECESSARY CONDITIONS

For convenience we repeat that Problem A is to find x^*, solving

$$\text{minimize } f(x)$$

subject to

$$g_i(x) \geq 0, \quad i = 1, \ldots, m, \tag{2.1}$$

$$h_j(x) = 0, \quad j = 1, \ldots, p. \tag{2.2}$$

Use will be made of the following lemma, which is stated here without proof. (For a proof see [61].)

Lemma 1 [Farkas [44]]. Let $\{a_k\}$, $k = 0, 1, \ldots, q$ be a set of $n \times 1$ vectors. A necessary and sufficient condition that there exist *non-negative* scalar values $\{t_k\}$, $k = 1, \ldots, q$ such that

$$a_0 = \sum_{k=1}^{q} t_k a_k \qquad (2.3)$$

is that for every vector z such that $z^T a_k \geq 0$, $k = 1, \ldots, q$, it follows that $z^T a_0 \geq 0$.

Definition. Next we define the Lagrangian function $\mathcal{L}(x, u, w)$ associated with Problem A as

$$\mathcal{L}(x, u, w) = f(x) - \sum_{i=1}^{m} u_i g_i(x) + \sum_{j=1}^{p} w_j h_j(x). \qquad (2.4)$$

We shall see that this function plays an important role in mathematical programming theory. This is particularly true for the duality theory associated with the convex programming problems of Chapter 6.

Suppose that x^* is a point satisfying the constraints of Problem A and the problem functions are differentiable. The following division of vectors in E^n is useful in establishing necessary conditions that must hold for x^* to be a local minimum. Let ∇f^* denote $\nabla f(x^*)$, and so on. Define

$$Z_1^* \equiv \{z \mid z^T \nabla g_i^* \geq 0 \ \text{(all } i \in B^*), \ z^T \nabla h_j^* = 0,$$
$$j = 1, \ldots, p \ \text{ and } \ z^T \nabla f^* \geq 0\},$$

$$Z_2^* \equiv \{z \mid z^T \nabla g_i^* \geq 0 \ \text{(all } i \in B^*), \ z^T \nabla h_j^* = 0,$$
$$j = 1, \ldots, p \ \text{ and } \ z^T \nabla f^* < 0\},$$

$$Z_3^* \equiv \{z \mid z^T \nabla g_i^* < 0 \ \text{(for at least one } i \in B^*) \ \text{ or }$$
$$z^T \nabla h_j^* \neq 0 \ \text{(for at least one } j)\},$$

where

$$B^* \equiv \{i \mid g_i(x^*) = 0\}.$$

Note that feasible perturbations from x^* must be contained in $Z_1^* \cup Z_2^*$. Among these $f(x)$ initially decreases along $z \in Z_2^*$ and initially increases or is constant along $z \in Z_1^*$. Thus if $Z_2^* \neq \emptyset$ we would not expect x^* to be a local minimum. We shall see that this expectation is correct, but only if we make a suitable regularity assumption. Clearly the sets are mutually disjoint, and any $z \in E^n$ belongs to one of the above three sets.

In terms of the sets Z_1^*, Z_2^*, and Z_3^* we are now ready to prove "first-order" necessary conditions for a local minimum. The term first-order refers to the assumption of once-continuous differentiability of the problem functions.

2.1 First-Order Necessary Conditions

Theorem 1 [Existence of Generalized Lagrange Multipliers]. If (a) x^* satisfies the constraints of Problem A, (b) the functions f, $\{g_i\}$, and $\{h_j\}$ are once differentiable, and (c) at x^* the set Z_2 is empty ($Z_2^* = \emptyset$), then it follows that there exist vectors u^* and w^* such that (x^*, u^*, w^*) satisfies

$$g_i(x) \geq 0, \qquad i = 1, \ldots, m, \tag{2.5}$$

$$h_j(x) = 0, \qquad j = 1, \ldots, p, \tag{2.6}$$

$$u_i g_i(x) = 0, \qquad i = 1, \ldots, m, \tag{2.7}$$

$$u_i \geq 0, \qquad i = 1, \ldots, m, \tag{2.8}$$

$$\nabla \mathcal{L}(x, u, w) = 0. \tag{2.9}$$

Proof. Assumption c implies that the hypotheses of Farkas' lemma are satisfied by the set $\{\nabla g_i^*\}$ (all $i \in B^*$), $\{\nabla h_j^*\}$ ($j = 1, \ldots, p$), $\{-\nabla h_j^*\}$ ($j = 1, \ldots, p$), and the vector ∇f^*. Then there exist scalars $u_i^* \geq 0$ (all $i \in B^*$), $y_j^* \geq 0$ ($j = 1, \ldots, p$), and $v_j^* \geq 0$ ($j = 1, \ldots, p$) such that

$$\nabla f^* = \sum_{i \in B^*} u_i^* \nabla g_i^* + \sum_{j=1}^{p} (y_j^* - v_j^*) \nabla h_j^*. \tag{2.10}$$

Let $u_i^* = 0$ for all $i \notin B^*$, and let $w_j^* = -y_j^* + v_j^*$ ($j = 1, \ldots, p$). Then (2.10) yields

$$\nabla f^* - \sum_{i=1}^{m} u_i^* \nabla g_i^* + \sum_{j=1}^{p} w_j^* \nabla h_j^* = 0$$

[which is just (2.9)], where

$$u_i^* \geq 0 \text{ if } g_i(x^*) = 0 \text{ and } u_i^* = 0 \text{ if } g_i(x^*) > 0 \qquad \text{Q.E.D.}$$

In applying Theorem 1 one must be able to determine if the set Z_2^* is empty. Clearly, assuming that the functions are differentiable, $Z_2^* = \emptyset$ is a necessary and sufficient condition for the existence of the vectors (u^*, w^*), called dual variables or generalized Lagrange multipliers.

Several conditions historically have been imposed to ensure that the set Z_2^* be empty at a local minimum. We give three of them in the following discussion. First we state a condition that may be required to hold at a candidate for a local solution of Problem A, a condition introduced by Kuhn and Tucker [77].

Definition [First-Order Constraint Qualification]. Let x^0 be a point satisfying (2.1) and (2.2) and assume that the functions $\{g_i\}$ and $\{h_j\}$ are once differentiable. Then the first-order constraint qualification holds at x^0 if for any nonzero vector z, such that $z^T \nabla g_i(x^0) \geq 0$ for all $i \in B_0 = \{i \mid g_i(x^0) = 0\}$ and $z^T \nabla h_j(x^0) = 0, j = 1, \ldots, p$, z is tangent to a once-differentiable arc, the arc emanating from x^0 and contained in the constraint region.

For purposes here an arc is a directed differentiable curve $\alpha(\theta)$ in E^n parameterized by the variable $\theta \geq 0$ in an interval $[0, \epsilon]$, where $\epsilon > 0$. We denote its tangent at $\theta = \eta$ by $D\alpha(\eta)$, and its second derivative by $D^2\alpha(\eta)$ if it is twice differentiable. Since curves can be rescaled by an arbitrary positive multiplication of the parameterizing variable, the use of the word tangent is usually applied to all vectors $\lambda D\alpha(\eta)$, where $\lambda > 0$. In the following development the restricted use of the term will cause no difficulty.

There are situations in which the first-order constraint qualification does not hold. Kuhn and Tucker [77] give a simple example involving the constraints

$$g_1(x) = (1 - x_1)^3 - x_2 \geq 0,$$
$$g_2(x) = x_1 \geq 0,$$
$$g_3(x) = x_2 \geq 0.$$

(No objective function need be given, since the first-order constraint qualification is concerned only with the constraints.) At the point $(1, 0)$ let Z be the set of vectors such that the hypotheses of the constraint qualification are satisfied. Thus we require $z \in Z$ to satisfy $(0, -1) z = -z_2 \geq 0$, and $(0, 1) z = z_2 \geq 0$. Then Z consists of all vectors of the form $(z_1, 0)$. When $z_1 < 0$, $(z_1, 0)$ is tangent to the arc $\alpha(\theta) = (1 + \theta z_1, 0)$, which is contained in the constraint region. When $z_1 > 0$, $(z_1, 0)$ cannot be tangent to an arc as required. Hence the first-order constraint qualification fails to hold for this example.

Corollary 1 [Kuhn-Tucker Necessity Theorem]. If the functions f, $\{g_i\}$, and $\{h_j\}$ are differentiable at x^* and if the first-order constraint qualification holds at x^*, then necessary conditions that x^* be a local minimum of Problem A are that there exist vectors u^* and w^* such that (x^*, u^*, w^*) satisfy (2.5–2.9).

Proof. We need to show that Z_2^* is empty. Consider any nonzero vector z such that $z^T \nabla g_i^* \geq 0$ for all $i \in B^*$ and $z^T \nabla h_j^* = 0, j = 1, \ldots, p$. If there is no such z, $Z_2^* = \emptyset$ and the corollary is proved. By the first-order constraint qualification, z is the tangent of a differentiable arc, emanating from x^* and contained in the constraint region. Let $\alpha(\theta)$ be that arc. The rate of change of f along $\alpha(\theta)$ at x^* is, since $x^* = \alpha(0)$ (making use of the chain rule), $Df[\alpha(0)] = [D\alpha(0)]^T \nabla f^* = z^T \nabla f^*$. Since $\alpha(\theta)$ is feasible and x^* is a local minimum, f must not decrease along $\alpha(\theta)$ starting from $\theta = 0$. Hence $z^T \nabla f^* \geq 0$. Thus $z \in Z_1^*$, and hence $Z_2^* = \emptyset$. Q.E.D.

The Kuhn-Tucker constraint qualification is one of the best known conditions implying that $Z_2^* = \emptyset$ [or, equivalently, implying the existence of the generalized Lagrange multipliers (u^*, w^*)]. A second condition that does this was given in [82] and is summarized in the following theorem.

2.1 First-Order Necessary Conditions

Corollary 2 [Interiority-Independence Necessity Theorem]. If (a) f, $g_1, \ldots, g_m, h_1, \ldots, h_p$ are differentiable functions of x; if (b) at a feasible point x^* there exists a vector s such that $s^T \nabla g_i^* > 0$ for all $i \in B^*$, $s^T \nabla h_j^* = 0$, $j = 1, \ldots, p$; and if (c) $\nabla h_1^*, \nabla h_2^*, \ldots, \nabla h_p^*$ are linearly independent vectors, then a necessary condition that x^* be a local minimum of (A) is that there exists (u^*, w^*) such that (x^*, u^*, w^*) satisfies (2.5–2.9).

Proof. Let z be any vector satisfying $z^T \nabla g_i^* \geq 0$, all $i \in B^*$, and $z^T \nabla h_j^* = 0$, $j = 1, \ldots, p$. We will show that $z \in Z_1^*$ and hence that Z_2^* is empty.

If $z \in Z_2^*$, then, because of our hypothesis, there is an $\epsilon > 0$ such that $y = z + \epsilon s$ satisfies $y^T \nabla g_i^* > 0$, all $i \in B^*$, $y^T \nabla h_j^* = 0, j = 1, \ldots, p$, and $y^T \nabla f^* < 0$. We construct an arc $\alpha(\theta)$ emanating from x^* as follows. Let $\alpha(0) = x^*$, and let

$$D\alpha(\theta) = P(\theta)r(y) = \{I - H[\alpha(\theta)]\{H^T[\alpha(\theta)]H[\alpha(\theta)]\}^{-1}H^T[\alpha(\theta)]\}r(y)$$

where

$$H[\alpha(\theta)] = \{\nabla h_1[\alpha(\theta)], \ldots, \nabla h_p[\alpha(\theta)]\}$$

and $r(y)$ is some vector such that $P(0)r(y) = y$. Such a vector exists since a necessary and sufficient condition that $H^T[\alpha(0)]y = 0$ is that

$$y = \{I - \{H^T[\alpha(0)]\}^{\#}H^T[\alpha(0)]\}r(y)$$

for some $r(y)$, where $A^{\#}$ is a pseudoinverse of A satisfying $AA^{\#}A = A$. It can be verified that $H[\alpha(0)]\{H^T[\alpha(0)]H[\alpha(0)]\}^{-1}$ is a pseudoinverse of $H^T[\alpha(0)]$ when the rows of $H^T[\alpha(0)]$ are independent (as assumed here).

Note that if there are no equality constraints we simply let $P(\theta) \equiv I$.

The above equation defines an arc, since in a neighborhood about x^* the inverse matrix [hence $P(\theta)$] exists and is continuous. Furthermore $D\alpha(0) = y \neq 0$.

Along this arc each $g_i[\alpha(\theta)]$ remains feasible, since

$$D\{g_i[\alpha(0)]\} = \nabla^T g_i^* y > 0.$$

For each $h_j[\alpha(\theta)]$, $h_j[\alpha(\theta)] = h_j[\alpha(0)] + \theta\, Dh_j[\alpha(\eta)]$ (for some $0 \leq \eta \leq \theta$) = $0 + \theta \nabla^T h_j[\alpha(\eta)]\{I - H[\alpha(\eta)]\{H^T[\alpha(\eta)]H[\alpha(\eta)]\}^{-1}H^T[\alpha(\eta)]\}r(y) = 0$. Thus $\alpha(\theta)$ is a feasible arc.

The rate of change of the objective function along $\alpha(\theta)$ at $\theta = 0$ is $Df[\alpha(0)] = \nabla^T f^* y < 0$. This contradicts the assumption that x^* is a local minimum. Hence $z \notin Z_2^*$ and Z_2^* is therefore empty.   Q.E.D.

A final condition implying the existence of (u^*, w^*) that will be developed here is the independence of the gradients of all the binding constraints at x^*. We prove this by proving a stronger result, that this condition implies the first-order constraint qualification.

Corollary 3 [Sufficient Condition for First-Order Constraint Qualification]. A sufficient condition that the first-order constraint qualification hold at a point x^0 satisfying the constraints of Problem A is that the gradients $\{\nabla g_i^0\}$, all $i \in B_0$, $\{\nabla h_j^0\}$, $j = 1, \ldots, p$ be linearly independent.

Proof. The proof is constructive. Let z be a nonzero vector satisfying $z^T \nabla g_i^0 \geq 0$ (all $i \in B_0$) and $z^T \nabla h_j^0 = 0$, $j = 1, \ldots, p$. If no such z exists the assertion is vacuously satisfied. Recall $B_0 \equiv \{i \mid g_i(x^0) = 0\}$. Let $E_0 = \{i \mid z^T \nabla g_i^0 = 0, i \in B^0\}$. Let G be the vector of constraints of all the g_i with $i \in E_0$ and all h_j. If G contains no elements it means that $z^T \nabla g_i^0 > 0$ for all $i \in B_0$, and there are no equality constraints in the problem. In this case let the arc constructed be simply the line emanating from $x^* = \alpha(0)$ with $D\alpha(\theta) \equiv z$ for θ included in $[0, \epsilon]$, $\epsilon > 0$.

By hypothesis z satisfies $z^T \nabla G = 0$. Thus z must be expressible in the form $z = [I - (\nabla^T G^0)^\# \nabla^T G^0] y(z)$ for some $y(z)$, where $\#$ indicates a pseudo-inverse of the matrix $\nabla^T G^0$. (See proof of Corollary 2 for definition of pseudo-inverse.) Since the rows of $\nabla^T G^0$ were assumed to be independent, one pseudoinverse of $\nabla^T G^0$ is given by $(\nabla^T G^0)^\# = \nabla G^0 [(\nabla^T G^0) \nabla G^0]^{-1}$.

The arc is constructed as follows. Let $\alpha(0) = x^0$ and

$$D\alpha(\theta) = \{I - \nabla G[\alpha(\theta)]\{\nabla^T G[\alpha(\theta)] \nabla G[\alpha(\theta)]\}^{-1} \nabla^T G[\alpha(\theta)]\} y(z) \quad (2.11)$$

for θ in $[0, \epsilon]$, $\epsilon > 0$. The inverse exists because in a neighborhood about x^0 the rows of $\nabla^T G[\alpha(\theta)]$ remain linearly independent. Because of continuity of the derivatives, (2.11) defines a curve. By construction, $D\alpha(0) = z$. All constraints where $i \in B_0$ and $z^T \nabla g_i^0 > 0$ must increase along $\alpha(\theta)$ and hence are feasible. For any components G_k of G,

$$G_k[\alpha(\theta)] = G_k[\alpha(0)] + \theta \, DG_k[\alpha(\eta)], \quad 0 \leq \eta \geq \theta$$
$$= 0 + \theta \, \nabla^T G_k[\alpha(\eta)] \, D\alpha(\eta)$$
$$= 0.$$

This completes the proof that $\alpha(\theta)$ is contained in the constraint region.

Q.E.D.

The following example shows that the first-order constraint qualification is not both *necessary* and sufficient for Z_2^* to be empty.

Example.

$$\text{minimize } x_2$$

subject to

$$(1 - x_1)^3 - x_2 \geq 0,$$
$$x_1 \geq 0,$$
$$x_2 \geq 0.$$

2.1 First-Order Necessary Conditions

These constraints do *not* satisfy the constraint qualification at $x^* = (1, 0)$, yet it is a local minimum; $Z_2^* = \emptyset$ and $(u_1^*, u_2^*, u_3^*) = (0, 0, 1)$, satisfying (2.5–2.9).

It is still an open question whether a weaker condition than the constraint qualification can be found to imply that $Z_2^* = \emptyset$ (the existence of finite multipliers). Certainly the criteria mentioned in Corollaries 2 and 3 are more useful because they can be tested. The constraint qualification is not a *useful* criterion for testing whether or not a candidate for a local minimum satisfies the necessary conditions. A weaker criterion, if found, will probably involve the objective function gradient ∇f^* as it is involved in defining the set Z_2^*.

In the sets already defined, we noted that $Z_2^* = \emptyset$ implies the existence of finite Lagrange multipliers. This condition is weaker than the constraint qualification, since the latter implies the former (Corollary 1) but not vice versa (see previous example). Both conditions suffer from the serious defect, with regard to numerical application, that they are not generally possible to verify computationally. The condition that $Z_2^* = \emptyset$ has the advantage, not only in being weaker, but in giving an explicit characterization of the desired condition directly in terms of the gradients of the problem functions at a solution. This condition is actually the fulfillment of the hypothesis of Farkas' lemma, which in turn is both necessary and sufficient for the existence of finite Lagrange multipliers. Because of the necessity of this condition, it is clear that it cannot be weakened.

The constraint qualification implies $Z_2^* = \emptyset$ and the consequent existence of finite multipliers, as we have seen. However, the qualification is not transparent, and though it offers an elegant criterion for assuming the desired result it has often unfortunately clouded the issue. It should properly be viewed as assuring that $Z_2^* = \emptyset$ so that Farkas' lemma can be directly invoked to assure finite Lagrange multipliers. There are other criteria for doing this, such as given in Corollary 2, and furthermore the constraint qualification is *not* a necessary condition for the finite multipliers to exist. It is certainly not a condition that need apply to a solution of even the most simple problems, as the example formulated by Kuhn and Tucker amply shows.

Very often in the literature the qualification is invoked with little explanation or comment and invariably the key consequence, namely that $Z_2^* = \emptyset$, is not emphasized. The result has been to mask the ensuing proofs and to give the unfortunate impression that the constraint qualification is somehow inextricably involved with the essential characterization of the solution of the mathematical programming problem. The latter impression has been reinforced by the development of various, often elaborate, conditions that assure that the qualification holds.

Another fact worth noting is that even the condition that $Z_2^* = \emptyset$ is *not* necessary at a solution. In view of what has been said, this would have to imply that the Lagrange multipliers are *not* finite. The very simple example provided by Kuhn and Tucker is a good illustration of this case. A wide class of problems gives rise to solutions that are thus characterized, even where the optimum is unique (as in the example mentioned). Further results in this class of problems should be possible and would be important. To this end, we note that the conditions required for the convergence of exterior point methods developed in this book (Chapter 4) admit the possibility of infinite multipliers and hence may provide a characterization of optimality under more general conditions.

The following corollary is an immediate result of Theorem 1 and Corollary 3.

Corollary 4 [First-Order Necessary Conditions for an Unconstrained Minimum]. A necessary condition that a differentiable function have an unconstrained local minimum at a point x^* is that

$$\nabla f(x^*) = 0. \tag{2.12}$$

Theorem 1 is a "first-order" characterization of local minima in that it involves the *first* partial derivatives of the problem functions. It does not, however, take into account the curvature of the functions, this being measured by the second partial derivatives. Curvature is the distinguishing characteristic between nonlinear and linear functions. The next example points out the need for curvature analysis to provide a finer characterization of situations in which the first-order necessary conditions fail to give complete information.

Example. We seek the values of the parameter $k > 0$ for which $(0, 0)$ is a local minimum of the problem

$$\text{minimize } (x_1 - 1)^2 + x_2^2$$

subject to

$$-x_1 + \frac{x_2^2}{k} \geq 0.$$

Since there is only one constraint, and its gradient is always nonzero, the first-order necessary conditions apply by virtue of Corollary 3. On the basis of (2.9) the following equation must be satisfied for some $u_1^* \geq 0$:

$$(-2, 0) - u_1^*(-1, 0) = (0, 0).$$

It follows that $u_1^* = 2$ satisfies this for all choices of $k > 0$. But for $k = 1$, $(0, 0)$ is not a local minimum and for $k = 4$ it is.

Thus the first-order necessary conditions indicate that (0, 0) is a candidate for a minimum, but do not yield any restrictions on the values of k. We shall see that second-order conditions give the additional information that (0, 0) is a minimum only for certain ranges of values of k.

2.2 SECOND-ORDER NECESSARY CONDITIONS

In [84] necessary conditions that apply to twice-differentiable functions were developed to take into account the curvature of the problem functions near a local minimum. They are summarized in Theorem 2. We shall require a further constraint qualification for this result.

Definition [Second-Order Constraint Qualification]. *Let x^0 be a point satisfying the constraints of (A), and assume that the functions $\{g_i\}$ and $\{h_j\}$ are twice differentiable.* The second-order constraint qualification holds at x^0 if the following is true. Let y be any nonzero vector such that $y^T \nabla g_i^0 = 0$ for all $i \in B_0 = \{i \mid g_i(x^0) = 0\}$, and such that $y^T \nabla h_j^0 = 0$, $j = 1, \ldots, p$. Then y is the tangent of an arc $\alpha(\theta)$, twice differentiable, along which $g_i[\alpha(\theta)] \equiv 0$ (all $i \in B_0$) and $h_j[\alpha(\theta)] \equiv 0$, $j = 1, \ldots, p$, where $\theta \in [0, \epsilon]$, $\epsilon > 0$.

Theorem 2 [Second-Order Necessary Conditions]. If the functions f, $\{g_i\}$, and $\{h_j\}$ are twice continuously differentiable and if the first- and second-order constraint qualifications hold at a point x^*, then necessary conditions that x^* be a local minimum to Problem A are that there exist vectors $u^* = (u_1^*, \ldots, u_m^*)^T$ and $w^* = (w_1^*, \ldots, w_p^*)^T$ such that (2.5 2.9) hold and such that for every vector y, where $y^T \nabla g_i^* = 0$, for all $i \in B^* = \{i \mid g_i(x^*) = 0\}$ and $y^T \nabla h_j^* = 0, j = 1, \ldots, p$, it follows that

$$y^T \nabla^2 \mathcal{L}(x^*, u^*, w^*) y \geq 0. \qquad (2.13)^*$$

Proof. (a) The first part of the theorem, the existence of vectors satisfying (2.5–2.9), follows from Corollary 1, since the first-order constraint qualification is assumed to hold.

(b) For the other conclusion, let y be any nonzero vector such that

$$y^T \nabla g_i^* = 0 \quad \text{for all} \quad i \in B^*, \qquad (2.14)$$

$$y^T \nabla h_j^* = 0, \quad j = 1, \ldots, p. \qquad (2.15)$$

If no such nonzero y exists, the conclusion follows vacuously. Otherwise let $\alpha(\theta)$ be the twice-differentiable arc guaranteed by the second-order constraint qualification, where $\alpha(0) = x^*$ and $D\alpha(0) = y$. Denote $D^2[\alpha(0)]$ by z.

---

* With a more general constraint qualification, (2.13) must hold for all y such that $y^T \nabla h_j^* = 0$, $j = 1, \ldots, p$, $y^T \nabla g_i^* \geq 0$, $i \in B^*$, and $y^T \nabla g_i^* = 0$ for all $i \in B^*$ with $u_i^* > 0$.

Then by the chain rule

$$D^2 g_i[\alpha(0)] = y^T \nabla^2 g_i^* y + z^T \nabla g_i^* = 0 \quad \text{for all} \quad i \in B^*, \quad (2.16)$$

$$D^2 h_j[\alpha(0)] = y^T \nabla^2 h_j^* y + z^T \nabla h_j^* = 0, \quad j = 1, \ldots, p, \quad (2.17)$$

because the g_i ($i \in B^*$) and h_j are identically equal to zero along $\alpha(\theta)$. Using the u^* and w^* given by Part a, and using (2.14) and (2.15),

$$Df[\alpha(0)] = y^T \nabla f^* = y^T \left(\sum_{i=1}^m u_i^* \nabla g_i^* - \sum_{j=1}^p w_j^* \nabla h_j^* \right) = 0.$$

Since x^* is a local minimum and $Df[\alpha(0)] = 0$, it follows that

$$D^2 f[\alpha(0)] \geq 0. \quad (2.18)$$

From this, (2.9), (2.16), and (2.17), it follows that

$$y^T [\nabla^2 f^* - \sum_{i=1}^m u_i^* \nabla^2 g_i^* + \sum_{j=1}^p w_j^* \nabla^2 h_j^*] y \geq 0. \quad \text{Q.E.D.}$$

In order to use these necessary conditions as criteria for determining if a point can be a local minimum, one must determine if the second order constraint qualification holds. The next theorem shows that independence of the gradients of the constraints equal to zero at x^* is sufficient for this, as it was also sufficient (see Corollary 3) for the first-order constraint qualification to hold.

Theorem 3 [Condition Sufficient for Second-Order Constraint Qualification]. Suppose that the functions $\{g_i\}$ and $\{h_j\}$ are twice differentiable. A sufficient condition that the second-order constraint qualification be satisfied at a point x^* satisfying the constraints of Problem A is that the vectors

$$\{\nabla g_i\} \text{ (all } i \in B^* = \{i \mid g_i(x^*) = 0\}), \{\nabla h_j^*\} \ (j = 1, \ldots, p)$$

be linearly independent.

Proof. We prove that the second-order constraint qualification holds by constructing the desired arc.

Let y be any nonzero vector satisfying (2.14) and (2.15). (If none exist, the second-order constraint qualification is trivially satisfied.) Let z be some vector such that

$$y^T \nabla^2 g_i^* y + z^T \nabla g_i^* = 0 \quad (\text{all } i \in B^*), \quad (2.19)$$

$$y^T \nabla^2 h_j^* y + z^T \nabla h_j^* = 0, \quad j = 1, \ldots, p. \quad (2.20)$$

Such a z exists because of the independence of the gradients. Assume that there are q indices in B^* and that inequality constraints are reordered so that

2.2 Second-Order Necessary Conditions

$g_1, \ldots, g_q$ are those constraints. Let $c = p + q$, which, by the assumption of linear independence and the existence of the nonzero y satisfying (2.14) and (2.15), must be less than n. Let e be the vector $[(p + q) \times 1]$ of constraints of all g_i ($i \in B_0$) and all h_j. Let $M_c(\theta)$ be the $(p + q) \times (p + q)$ matrix whose i,jth element is $\partial e_i[\alpha(\theta)]/\partial x_j$, where $i = 1, \ldots, p + q$, $j = 1, \ldots, p + q$. Let $M_{cn}(\theta)$ be the $(p + q) \times [n - (p + q)]$ matrix whose i,jth element is $\partial e_i[\alpha(\theta)]/\partial x_j$, where $i = 1, \ldots, p + q, j = p + q + 1, \ldots, n$. Let $d(\theta)$ be a $(p + q) \times 1$ vector whose ith element is

$$[D\alpha(\theta)]^T \nabla^2 e_i[\alpha(\theta)] \, D\alpha(\theta)$$

for $i = 1, \ldots, p + q$. Let z_{cn} be the $[n - (p + q)] \times 1$ vector whose ith component is $z_i, i = p + q + 1, \ldots, n$. We can assume without loss of generality that $M_c(0)$ has rank $(p + q)$.

The arc is constructed as follows. Let $\alpha(0) = x^*$, $D\alpha(0) = y$, and

$$D^2\alpha(\theta) = \begin{bmatrix} [M_c(\theta)]^{-1}[-M_{cn}(\theta)z_{cn} - d(\theta)] \\ z_{cn} \end{bmatrix}. \tag{2.21}$$

This defines an arc, since the right-hand side is continuous in a neighborhood about x^*. It is a nontrivial arc, since y was assumed not equal to zero. That $D\alpha(0) = z$ follows by solving (2.19) and (2.20) for $(z_1, \ldots, z_{p+q})$ in terms of z_{cn} and noting that this agrees with (2.21) at $\theta = 0$.

Expanding c_i (where e_i is any component of e) in a Taylor's series yields

$$e_i[\alpha(\theta)] = e_i[\alpha(0)] + \theta \, De_i[\alpha(0)] + \frac{\theta^2}{2} D^2 e_i[\alpha(\varphi)],$$

where $0 \leq \varphi \leq \theta$, in a neighborhood about $\alpha(0) = x^*$. But $e_i[\alpha(0)] = 0$, and $De_i[\alpha(0)] = \nabla^T e_i(x^*)y = 0$ [using the chain rule and (2.14) and (2.15)]. Using the chain rule again,

$$D^2 e_i[\alpha(\varphi)] = d^i(\varphi) + [M_c^i(\varphi), M_{cn}^i(\varphi)] \begin{bmatrix} \{M_c(\varphi)\}^{-1}[-M_{cn}(\varphi)z_{cn} - d(\varphi)] \\ z_{cn} \end{bmatrix}$$

$$= 0. \qquad \text{Q.E.D.}$$

The following example illustrates that the first-order constraint qualification can be satisfied and the second-order constraint qualification can fail to hold.

minimize x_2

subject to

$$g_1 = -x_1^9 + x_2^3 \geq 0,$$
$$g_2 = x_1^9 + x_2^3 \geq 0,$$
$$g_3 = x_1^2 + (x_2 + 1)^2 - 1 \geq 0.$$

The solution is at $x^* = (0, 0)^T$. Now $\nabla g_1^* = \nabla g_2^* = (0, 0)^T$, $\nabla g_3^* = (0, 2)^T$. Since all three constraints are equal to zero at x^*, $B^* = \{1, 2, 3\}$. Any vector y such that $y^T \nabla g_i^* \geq 0$, all $i \in B^*$, must be of the form $(y_1, y_2^2)^T$. Clearly any such vector is tangent to an arc pointing into the constraint region. Thus the first-order constraint qualification is satisfied.

Any vector y to be considered for the second-order constraint qualification is of the form $(y_1, 0)^T$ (where $y_1 \neq 0$ since y must be nonzero). Since there is *no* arc along which g_1, g_2, and g_3 remain equal to zero, the second-order constraint qualification fails to hold.

Note that the first-order Lagrange conditions are satisfied only by $(u_1^*, u_2^*, \frac{1}{2})$, where u_1^* and u_2^* are any non-negative scalars. However,

$$\nabla^2 L(x^*, u^*) = \begin{bmatrix} -1 & 0 \\ 0 & -1 \end{bmatrix}$$ is negative definite, and the second-order necessary conditions do *not* hold.

That the second-order constraint qualification does not imply the first can be seen in the following example.

$$g_1 = -x_1^2 - (x_2 - 1)^2 + 1 \geq 0,$$

$$g_2 = -x_1^2 - (x_2 + 1)^2 + 1 \geq 0,$$

$$g_3 = x_1 \geq 0.$$

The point $(0, 0)^T$ is the solution to any problem with these three constraints. Their gradients are $(0, 2)^T$, $(0, -2)^T$, and $(1, 0)^T$. The second-order constraint qualification *is* satisfied because there *are* no vectors orthogonal to all three gradients. The first-order qualification is not, since there are no arcs pointing into the region of feasibility (which is a single point). There are vectors y giving non-negative inner products with all the constraint gradients at $(0, 0)^T$, for example, $y = (2, 0)^T$.

A corollary of the last two theorems is the following well-known result.

Corollary 5. *Necessary conditions that a twice continuously differentiable function have a local unconstrained minimum at a point x^* are that*

$$\nabla f(x^*) = 0 \tag{2.22}$$

and

$$y^T \nabla^2 f(x^*) y \geq 0 \quad \text{for all} \quad y. \tag{2.23}$$

Definition. *A symmetric matrix A is said to be* positive semidefinite *if, for every vector y, $y^T A y \geq 0$.* Thus (2.23) is equivalent to saying

$$\nabla^2 f(x^*) \quad \text{is a positive semidefinite matrix.} \tag{2.24}$$

2.2 Second-Order Necessary Conditions

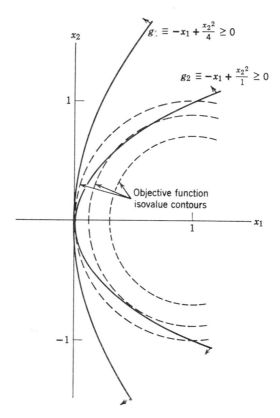

Figure 2 Failure of first-order conditions. $(0, 0)$ *solves* minimize $f(x)$ subject to $g_1(x) \geq 0$. $(0, 0)$ *does not solve* minimize $f(x)$ subject to $g_2(x) \geq 0$.

In the example of Figure 2, at $x^* = (0, 0)$, the matrix

$$[\nabla^2 f - \sum u_i \nabla^2 g_i] = \begin{bmatrix} 2 & 0 \\ 0 & 2 - \dfrac{4}{k} \end{bmatrix}. \tag{2.25}$$

Now since $\nabla g_1^* = (-1, 0)^T$, we need only consider vectors of the form $(0, y_2)$ in checking the second-order necessary conditions. Pre- and postmultiplying the matrix in (2.25) by $(0, y_2)$ yields $y_2^2 [2 - 4/k]$. By Theorem 2, for $k < 2$, $(0, 0)$ *is not* a local minimum. For $k \geq 2$ the second-order necessary conditions are satisfied.

It still remains to show that for certain values of k $(0, 0)^T$ *is* a local minimum. The following development, which appeared in [84] and [71], establishes a test of sufficiency, involving second partial derivatives, that guarantees that a point will be a local minimum.

2.3 SECOND-ORDER SUFFICIENCY CONDITIONS

The following conditions constitute an attempt to add as little as possible to the necessary conditions of Theorem 2 to provide sufficient conditions that a point be an isolated local minimum.

Theorem 4 [Second-Order Sufficiency Conditions]. Sufficient conditions that a point x^* be an *isolated* (unique locally) local minimum of Problem A, where f, $\{g_i\}$, and $\{h_j\}$ are twice-differentiable functions, are that there exist vectors u^* and w^* such that (x^*, u^*, w^*) satisfies

$$g_i(x) \geq 0, \quad i = 1, \ldots, m, \tag{2.26}$$

$$h_j(x) = 0, \quad j = 1, \ldots, p, \tag{2.27}$$

$$u_i g_i(x) = 0, \quad i = 1, \ldots, m, \tag{2.28}$$

$$u_i \geq 0, \quad i = 1, \ldots, m, \tag{2.29}$$

$$\nabla \mathcal{L}(x, u, w) = 0, \tag{2.30}$$

and for every nonzero vector y satisfying $y^T \nabla g_i^* = 0$ for all $i \in D^* = \{i \mid u_i^* > 0\}$, $y^T \nabla g_i^* \geq 0$, $i \in B^* - D^*$, and $y^T \nabla h_j^* = 0$, $j = 1, \ldots, p$, it follows that

$$y^T [\nabla^2 \mathcal{L}(x, u, w)] y > 0. \tag{2.31}$$

Proof. Assume that x^* is not an isolated local minimum. Then there exists a sequence of points $\{y^k\}$ where $\lim_{k \to \infty} y^k = x^*$ such that (a) each y^k is feasible and (b) $f(y^k) \leq f(x^*)$. We can rewrite each y^k as $x^* + \delta^k s^k$, where $\|s^k\| = 1$ and $\delta^k > 0$ for each k. We consider any limit point of the sequence $\{\delta^k, s^k\}$. Clearly any such limit point is of the form $(0, \bar{s})$, where $\|\bar{s}\| = 1$. By Property a,

$$g_i(y^k) - g_i(x^*) \geq 0 \quad (\text{all } i \in B^*),$$
$$h_j(y^k) - h_j(x^*) = 0, \quad j = 1, \ldots, p,$$

and by Property b,

$$f(y^k) - f(x^*) \leq 0.$$

Dividing each equation above by δ^k, and taking the limit as $k \to \infty$ (using that subsequence converging to $\bar{s}$), we have, by the assumed differentiability properties that

$$\nabla^T g_i^* \bar{s} \geq 0 \quad (\text{all } i \in B^*), \tag{2.32}$$

$$\nabla^T h_j^* \bar{s} = 0, \quad j = 1, \ldots, p, \tag{2.33}$$

and

$$\nabla^T f^* \bar{s} \leq 0. \tag{2.34}$$

2.3 Second-Order Sufficiency Conditions 31

We have two cases to consider and show that a contradiction arises from each of them.

(1) For the unit vector $\bar{s}$, $\nabla^T g_i^* \bar{s} > 0$ for at least one $i \in D^*$. But this, coupled with (2.32), (2.33), (2.34), (2.28), (2.29), and (2.30), means that

$$0 \geq \nabla^T f^* \bar{s} = \sum_{i \in D^*} u_i^* \nabla^T g_i^* \bar{s} - \sum_{j=1}^{p} w_j^* \nabla^T h_j^* \bar{s} > 0,$$

which is a contradiction.

(2) For the unit vector $\bar{s}$,

$$\nabla^T g_i^* \bar{s} = 0 \quad \text{for all} \quad i \in D^*, \quad \text{or} \quad D^* \text{ is empty}.$$

Then, using Taylor's expansion [and (a) and (b)],

$$0 \leq g_i(y^k) = g_i(x^* + \delta^k s^k)$$

$$= g_i(x^*) + \delta^k \nabla^T g_i^* s^k + \frac{(\delta^k)^2}{2}(s^k)^T [\nabla^2 g_i(\eta_i)] s^k \quad \text{(all } i \in B^*), \tag{2.35}$$

$$0 = h_j(y^k) = h_j(x^*) + \delta^k \nabla^T h_j^* s^k + \frac{(\delta^k)^2}{2}(s^k)^T [\nabla^2 h_j(\eta_j)] s^k, \quad j = 1, \ldots, p,$$

and (2.36)

$$0 \geq f(y^k) - f(x^*)$$

$$= \delta^k \nabla^T f^* s^k + \frac{(\delta^k)^2}{2}(s^k)^T [\nabla^2 f(\eta^0)] s^k. \tag{2.37}$$

Multiplying each term in (2.35) by $(-u_i^*)$, each term in (2.36) by w_j^*, and adding the results to (2.37) yields

$$0 \geq \delta^k \left[\nabla f^* - \sum_{i \in D^*} u_i^* \nabla g_i^* + \sum_{j=1}^{p} w_j^* \nabla h_j^* \right]^T s^k$$

$$+ \frac{(\delta^k)^2}{2}(s^k)^T \left[\nabla^2 f(\eta^0) - \sum_{i \in D^*} u_i^* \nabla^2 g_i(\eta_i) + \sum_{j=1}^{p} w_j^* \nabla^2 h_j(\eta_j) \right] s^k. \tag{2.38}$$

Using (2.30), dividing (2.38) by $(\delta^k)^2/2$, and taking the limit as $k \to \infty$ yields, because of Assumption 2, a statement contradicting (2.31). Q.E.D.

The following corollary results.

Corollary 6. Sufficient conditions that a point x^* be an isolated local unconstrained minimum of the twice-differentiable function f are that

$$\nabla f(x^*) = 0 \tag{2.39}$$

and
$$y^T \nabla^2 f(x^*) y > 0 \quad \text{for all nonzero} \quad y. \tag{2.40}$$

Definition. *A symmetric matrix A is positive definite if, for every nonzero* y, $y^T A y > 0$. Thus (2.40) is equivalent to

$$\nabla^2 f(x^*) \quad \text{is a positive definite matrix.} \tag{2.41}$$

Applying Theorem 4 to the example of Figure 2, for $k > 2$, it follows that $(0, 0)^T$ *is* a local minimum. Thus, using the second-order conditions, all cases are disposed of except $k = 2$. Here the necessary conditions are satisfied, but not the sufficient ones.

Discussion of Sufficiency Conditions

For general programming problems it does not appear to be possible, by utilizing information only at a point, to state conditions that use up to second partial derivatives, which are simultaneously necessary and sufficient for the point to be a local minimum. This can be seen by examining two trivial constrained optimization problems: (a) minimize x_1^4 (no constraints) and (b) minimize $x_1^3 + x_1^4$ (no constraints). For Problem a $x_1^* = 0$ is an isolated minimum but the sufficient conditions of Theorem 4 are not satisfied. For Problem b $x_1^* = 0$ satisfies the necessary conditions but not the sufficient ones, yet is *not* a local minimum.

It is important to realize that Theorem 4 states *two* things: conditions for which x^* is a local minimum, and also the fact that x^* is an isolated local minimum; that is, x^* is unique locally.

The isolated or "unique" property of the sufficient conditions of Theorem 4 are closely related to the conditions often required for a solution of a set of n simultaneous equations in n unknowns to be unique—that the Jacobian of the set of equations be nonzero. Another way of regarding this is that the first-order necessary conditions are uniquely satisfied.

Corollary 7 [Jacobian Condition Implying Sufficiency]. If $f, g_1, \ldots, g_m$, $h_1, \ldots, h_p$ are twice-differentiable functions of x and if, at a point x^*, (a) conditions (2.5–2.9) and (2.13) hold and if (b) the Jacobian of (2.6), (2.7), and (2.9) with respect to (x, u, w) does not vanish at (x^*, u^*, w^*), then the sufficiency conditions of Theorem 4 are satisfied at x^*.

Proof. The proof is a result of the material in Theorem 14, Section 5.2, and will not be given here.

Typically Conditions a and b of the corollary will hold at a local minimum. By assuming them it is possible to prove strong results concerning the convergence of algorithms, sensitivity analysis, and so on (see Chapter 5 and

2.3 Second-Order Sufficiency Conditions

Section 2.4 for examples). Before we leave this subject let us examine the implications of Theorem 4 for a linear programming problem.

In the linear problem the constraints and the objective function have matrices of second partial derivatives that are everywhere equal to zero. Immediately this implies that (2.31) can only hold in the logical sense, that is, there are *no* nonzero vectors orthogonal to all the indicated constraint gradients. This means that there are at least n independent gradients in the indicated set, and hence the solution point is uniquely given by an extreme point of the convex polyhedron defined by the constraints.

A final result of this section, proved in [48], which will be useful later, is a statement of conditions sufficient for x^* to be a local (not necessarily isolated) minimum of Problem A.

Theorem 5 [Neighborhood Sufficiency Theorem]. Let

$$Y = \{y \mid y^T \nabla g_i^* = 0 \ (i \in D^*), y^T \nabla g_i^* \geq 0,$$
$$i \in B^* - D^*, y^T \nabla h_j^* = 0 \quad (\text{all } j), \|y\| = 1\}.$$

Define $Z(\epsilon, \delta) \equiv \{z \mid \|z - y\| \leq \epsilon \text{ for some } y \in Y, x^* + \delta_z z \text{ is feasible, and } \|z\| = 1, \text{ where } 0 < \delta_z < \delta, \epsilon > 0\}$.

If (a) (2.26–2.30) are satisfied at (x^*, u^*, w^*) and (b) there exists a $\bar{Z} = Z(\bar{\epsilon}, \bar{\delta})$ such that for all $z \in \bar{Z}$

$$z^T \nabla^2 \mathcal{L}(x^* + \lambda \delta_z z, u^*, w^*) z \geq 0 \tag{2.42}$$

for all λ, such that $0 \leq \lambda \leq 1$, then x^* is a local minimum of Problem A.

Proof. The proof proceeds in a manner similar to the proof of Theorem 4. If the theorem were not true there would exist a sequence of feasible points y^k such that $f(y^k) < f(x^*)$, all k, and $\lim_{k \to \infty} y^k = x^*$.

Let $y^k = x^* + \delta^k s^k$, where $\|s^k\| = 1$ and $\delta^k > 0$. Let $\{s^k\}$ denote a converging subsequence, i.e. assume $\lim_{k \to \infty} (\delta^k, s^k) = (0, \bar{s})$, where $\|\bar{s}\| = 1$.

Suppose that $\nabla^T g_i^* \bar{s} > 0$ for at least one $i \in D^* \equiv \{i \mid u_i^* > 0\}$. Then the same contradiction arises as arose in the first part of Theorem 4.

Suppose that $\nabla^T g_i^* \bar{s} = 0$ for all $i \in D^*$, or suppose that D^* is empty. Then $\bar{s} \in Y$. A Taylor expansion yields

$$\mathcal{L}(y^k, u^*, w^*) = \mathcal{L}(x^*, u^*, w^*) + \delta^k (s^k)^T \nabla \mathcal{L}(x^*, u^*, w^*)$$
$$+ \frac{(\delta_k)^2}{2} (s_k)^T \nabla^2 \mathcal{L}(\eta^k, u^*, w^*) s^k, \tag{2.43}$$

where $\eta^k = x^* + \lambda \delta^k s^k, 0 \leq \lambda \leq 1$.

Using all our assumptions (2.43) reduces to

$$(s^k)^T \nabla^2 \mathcal{L}(\eta^k, u^*, w^*) s^k < 0 \tag{2.44}$$

for all k.

Clearly, for k large enough, $s^k \in Z(\bar{\epsilon}, \bar{\delta})$. By our hypothesis, then, for k large enough,
$$(s^k)^T \nabla^2 \mathcal{L}(\eta^k, u^*, w^*) s^k \geq 0.$$
This contradicts (2.44). Q.E.D.

There are instances in which this sufficiency theorem can be applied because the behavior of the problem functions is known in a neighborhood about the minimum. For convex programming this is the case (see Chapter 6).

(We note that this theorem can be used to show that, for $k = 2$, $(0, 0)^T$ must be a local solution to the example of Figure 2. Thus all possible cases are taken care of for that example.)

2.4 SENSITIVITY ANALYSIS IN NONLINEAR PROGRAMMING

An important question that often arises in the utilization of mathematical programming algorithms is how much the optimum changes when changes are made in the constraints and the objective function. Stated formally, a general parametric programming problem is

$$\text{minimize } f(x) + \epsilon_0 a_0(x)$$

subject to

$$g_i(x) + \epsilon_i b_i(b) \geq 0, \quad i = 1, \ldots, m, \quad \text{P}(\epsilon) \quad (2.45)$$

$$h_j(x) + \epsilon_{j+m} c_j(x) = 0, \quad j = 1, \ldots, p. \quad (2.46)$$

Here $a_0(x)$, $\{b_i(x)\}$, and $\{c_j(x)\}$ are scalar functions of x, and $\epsilon = (\epsilon_i, \ldots, \epsilon_{p+m})$. The question can now be stated: What is known about $[x(\epsilon) - x^*]$, where $x(\epsilon)$ is a local minimum of P(ϵ) near $x^* = x(0)$?

An answer to this question is available if the appropriate assumptions are made about conditions holding at x^*. If, along with the second-order sufficiency conditions of Theorem 4, some regularity conditions are assumed to hold at x^*, then a result concerning perturbation of the optimum can be summarized as follows.

Theorem 6 [Perturbation of optimum]. If (a) the functions of P(ϵ) are twice differentiable, (b) the sufficiency conditions (2.26–2.31) of Theorem 4 hold at x^*, (c) the gradients $\{\nabla g_i^*\}$ (all $i \in B^*$), $\{\nabla h_j^*\}$, $j = 1, \ldots, p$, are linearly independent, and (d) strict complementarity holds in (2.28) [that is, $u_i^* > 0$ when $g_i(x^*) = 0$], then

(i) the multipliers $\{u_i^*\}$ and $\{w_j^*\}$ are unique; (ii) there exists a unique differentiable function $[x(\epsilon), u(\epsilon), w(\epsilon)]$, where $x(\epsilon)$ is a local minimum of Problem P(ϵ), and $[u(\epsilon), w(\epsilon)]$ are the multipliers associated with it, where $x(0), u(0), w(0) = [x^*, u^*, w^*]$,

2.4 Sensitivity Analysis in Nonlinear Programming

(iii) a differential approximation to

$$\begin{bmatrix} x(\epsilon) - x^* \\ u(\epsilon) - u^* \\ w(\epsilon) - w^* \end{bmatrix}$$

is given by

$$\begin{bmatrix} \nabla^2 \mathcal{L}^* & -G^* & H^* \\ U^*(G^*)^T & \text{diag}(g_i^*) & 0 \\ (H^*)^T & 0 & 0 \end{bmatrix}^{-1} \begin{bmatrix} -\nabla a_0^* & B^*U^* & -C^*W^* \\ 0 & \text{diag}(-u_i^* b_i^*) & 0 \\ 0 & 0 & \text{diag}(-c_j^*) \end{bmatrix} \epsilon, \quad (2.47)$$

where $U^* = \text{diag}(u_i^*)$, $G^* = (\nabla g_1^*, \ldots, \nabla g_m^*)$, $H^* = (\nabla h_1^*, \ldots, \nabla h_p^*)$ $B^* = (\nabla b_1^*, \ldots, \nabla b_m^*)$, $W^* = \text{diag}(w_j^*)$, and $C^* = (\nabla c_1^*, \ldots, \nabla c_p^*)$; and (iv) the change in the optimum value of the objective function $f[x(\epsilon)] - f(x^*)$ is approximated by

$$\epsilon_0 a_0^* - \sum_{i=1}^{m} \epsilon_i u_i^* b_i^* + \sum_{j=1}^{p} \epsilon_{j+m} w_j^* c_j^*. \quad (2.48)$$

Proof. Because of Assumption b, the following set of equations is satisfied at $(x, u, w) = (x^*, u^*, w^*)$, $\epsilon = 0$:

$$\nabla f + \epsilon_0 \nabla a_0 - \sum u_i [\nabla g_i + \epsilon_i \nabla b_i] + \sum w_j [\nabla h_j + \epsilon_{j+m} \nabla c_j] = 0, \quad (2.49)$$

$$u_i[g_i(x) + \epsilon_i b_i(x)] = 0, \quad i = 1, \ldots, m, \quad (2.50)$$

$$h_j(x) + \epsilon_{j+m} c_j(x) = 0, \quad j = 1, \ldots, p. \quad (2.51)$$

[Then Part i, uniqueness of the multipliers, follows from the independence assumptions (c) and (2.49).]

This is a system of $(n + m + p)$ equations in $(n + 2m + 2p + 1)$ unknowns. The Jacobian matrix of this system with respect to (x, u, w) at (x^*, u^*, w^*), $\epsilon = 0$, is

$$M^* = \begin{bmatrix} \nabla^2 \mathcal{L}^* & -G^* & H^* \\ U^*(G^*)^T & \text{diag}(g_i^*) & 0 \\ (H^*)^T & 0 & 0 \end{bmatrix}, \quad (2.52)$$

where the elements of M^* are those defined after (2.47).

It is possible to show (see Theorem 14, Section 5.2) that under Assumptions a–d this matrix has an inverse. From the implicit function theorem [102, page 181] a unique differentiable vector function $[x(\epsilon), u(\epsilon), w(\epsilon)]$ in a neighborhood about $\epsilon = 0$ is obtained, satisfying (2.49–2.51). That the $x(\epsilon)$

is a solution to the parametric problem (P) is seen as follows, by proving that the sufficiency conditions (2.26–2.31) are satisfied by $[x(\epsilon), u(\epsilon), w(\epsilon)]$.

Conditions (2.30), (2.28), and (2.27) follow from (2.49), (2.50), and (2.51). (All g_i where $g_i(x^*) > 0$ are obviously well satisfied for ϵ small.)

That (2.26) holds for $i \in B^*$ (in fact $g_i[x(\epsilon)] + \epsilon_i b_i[x(\epsilon)] = 0$) follows by dividing (2.50) by $u_i(\epsilon)$ for all $i \in B^*$. This is possible because of the assumption that $u_i^* > 0$ for $i \in B^*$. The same arguments apply to prove that $u_i(\epsilon) > 0$, all $i \in B^*$, $u_i(\epsilon) = 0$, all $i \notin B^*$. Hence (2.29) is satisfied.

If there exists a vector function $z(\epsilon)$ of length one such for all ϵ small, $z(\epsilon)^T \nabla g_i[x(\epsilon)] = 0$, $i \in B^*$, $z(\epsilon)^T \nabla h_j[x(\epsilon)] = 0$, $j = 1, \ldots, p$, and $z(\epsilon)^T \nabla^2 \mathscr{L}[x(\epsilon), u(\epsilon), w(\epsilon)]z(\epsilon) \leq 0$, then with the results proved above, Assumption c implies there is a nonzero vector z^* violating (2.31). Therefore for ϵ small, the second order sufficiency conditions are satisfied at $x(\epsilon)$.

It is possible to be explicit about the derivative of the differentiable functions whose existence was assured by the implicit function theorem. Treating (2.49), (2.50), and (2.51) as identities in ϵ, differentiating, evaluating at (x^*, u^*, w^*), $\epsilon = 0$, and rearranging yields

$$\begin{bmatrix} \dfrac{dx(0)}{d\epsilon} \\[6pt] \dfrac{du(0)}{d\epsilon} \\[6pt] \dfrac{dw(0)}{d\epsilon} \end{bmatrix} = \begin{bmatrix} \nabla^2 \mathscr{L}^* & -G^* & H^* \\ U^*(G^*)^T & \operatorname{diag}(g_i^*) & 0 \\ (H^*)^T & 0 & 0 \end{bmatrix}^{-1} \times \begin{bmatrix} -\nabla a_0^* & B^*U^* & -C^*W^* \\ 0 & \operatorname{diag}(-u_i^* b_i^*) & 0 \\ 0 & 0 & \operatorname{diag}(-c_j^*) \end{bmatrix}. \quad (2.53)$$

Part iii follows directly from (2.53). Let $\dot{x}$ denote $dx(0)/d\epsilon$, and so on.

To prove Part iv we first use the chain rule and (2.49) to obtain

$$\frac{df[x(0)]}{d\epsilon} = \nabla^T f^* \dot{x} + \partial f/\partial \epsilon$$

$$= (\sum u_i^* \nabla g_i^* - \sum w_j^* \nabla h_j^*)^T \dot{x} + (a_0^*, 0, \ldots, 0) \quad (2.54)$$

Using all the properties proved above, differentiating (2.50) and (2.51) yields

$$u_i^* \nabla^T g_i^* \dot{x} = (0, 0, \ldots, 0, -u_i^* b_i^*, 0, \ldots 0), \quad i = 1, \ldots, m,$$

$$\nabla^T h_j^* \dot{x} = (0, 0, \ldots, 0, -c_j^*, 0, \ldots, 0), \quad j = 1, \ldots, p.$$

2.4 Sensitivity Analysis in Nonlinear Programming

Using these and substituting in (2.54) yields

$$\frac{df[x(0)]}{d\epsilon} = (a_0^*, -u_1^* b_1^*, \ldots, -u_m^* b_m^*, w_1^* c_1^*, \ldots, w_p^* c_p^*). \quad (2.55)$$

The differential approximation using (2.55) yields (2.48).   Q.E.D.

Equation 2.55 explains why the multiplier values (u^*, w^*) are often called "shadow prices." In effect they represent the rate of change of the optimum objective function value with respect to changes in the constraints.

The results implied by (2.53) are much more general than that given by (2.55). What (2.53) yields is the rate of change of the optimal allocation vector *and* of the shadow prices as functions of changes in the constraints.

Use of this sensitivity analysis is illustrated in the next example.

Example.   The problem in E^2 is

$$\text{minimize } -x_1 - x_2$$

subject to

$$-x_1 - x_2^2 + \epsilon_1 \geq 0.$$

Using the first-order necessary conditions the general solution is

$$x_1(\epsilon) = -\tfrac{1}{4} + \epsilon_1,$$

$$x_2(\epsilon) = \tfrac{1}{2},$$

$$u_1(\epsilon) = 1.$$

At $\epsilon_1 = 0$, using (2.53),

$$\begin{pmatrix} \frac{dx(0)}{d\epsilon_1} \\ \frac{du(0)}{d\epsilon_1} \end{pmatrix} = \begin{pmatrix} 0 & 0 & 1 \\ 0 & 2 & 1 \\ -1 & -1 & 0 \end{pmatrix}^{-1} \begin{pmatrix} 0 \\ 0 \\ -1 \end{pmatrix}$$

$$= \begin{pmatrix} \tfrac{1}{2} & -\tfrac{1}{2} & -1 \\ -\tfrac{1}{2} & \tfrac{1}{2} & 0 \\ 1 & 0 & 0 \end{pmatrix} \begin{pmatrix} 0 \\ 0 \\ -1 \end{pmatrix} = \begin{pmatrix} 1 \\ 0 \\ 0 \end{pmatrix}.$$

Thus the agreement in this example is exact for all values of ϵ_1. The solution for all values of ϵ_1 is completely characterized by (2.53) using the information available at one solution.

2.5 HISTORICAL REMARKS

Various results related to the theory developed in this chapter can be found scattered throughout the literature. For example, for the problem minimize $f(x)$ subject to $h_j(x) = 0, j = 1, \ldots, p$, assuming the $\{\nabla h_j(x),\text{ all } j\}$ linearly independent, Carathéodory [19] proved the necessary conditions for a minimum and the sufficient conditions for an isolated minimum. These are particular cases of the results obtained in this chapter.

Related theoretical results for both constrained and unconstrained minimization can be found in the extensive treatment given by Hancock [69], which first appeared in 1917.

For the problem of maximizing a quadratic indefinite form subject to linear constraints, similar results are given by Ritter [97].

Additional developments may be found in Hestenes [71, Chapter 1], and in the works of Fritz John, William Karush, Louis L. Pennisi, and R. Pallu de la Barrière (see Additional References).

ADDITIONAL REFERENCES

John, F., "Extremum Problems with Inequalities as Subsidiary Conditions," *Studies and Essays*, New York, 1948, pp. 187–204.

Karush, W., "Minima of Functions of Several Variables with Inequalities as Side Conditions," University of Chicago Master's Thesis, 1939.

Pallu de la Barrière, R., "Compléments a la Théorie des Multiplicateurs en Programmation Non-linèaire," *Revue Francaise de Recherche Operationelle*, **7**, 27, 163–180 (1963).

Pennisi, L., "An Indirect Sufficiency Proof for the Problem of Lagrange with Differential Inequalities as Added Side Conditions," *Trans. of the Am. Math. Soc.*, **74**: 177–198 (1953).

3

Interior Point Unconstrained Minimization Techniques

3.1 INTRODUCTION—DERIVATION OF ALGORITHMS FROM SUFFICIENCY CONDITIONS FOR CONSTRAINED MINIMA

A number of algorithms for solving nonlinear programming problems can be derived from the sufficiency conditions given in Section 2.3 for a point to be a local constrained minimum. These algorithms, including the unconstrained minimization algorithms discussed in this section, can be viewed as attempts to satisfy these sufficiency conditions. In many cases the algorithms were originally developed from geometric analysis rather than the algebraic way described here. The intuitive reasoning behind them will be given in Section 3.2.

The present chapter is concerned with the problem when there are no equality constraints, that is, problems of the form

$$\text{minimize} f(x) \tag{B}$$

subject to
$$g_i(x) \geq 0, \quad i = 1, \ldots, m.$$

We also assume, temporarily, that about a local minimum x^* of Problem B there is a neighborhood in which the constraints can be strictly satisfied; that is, there exist points x^0 such that $g_i(x^0) > 0$, $i = 1, \ldots, m$. Assume also that strict complementarity holds in (2.28); that is, $u_i^* > 0$ if $g_i(x^*) = 0$.

Proceeding formally, we consider a perturbation of the conditions (2.26–2.31) sufficient for x^* to be a local minimum. Suppose that the following conditions hold at some point $[x(r), u(r)]$ near (x^*, u^*) for r small:

$$g_i(x) > 0, \quad i = 1, \ldots, m, \tag{3.1}$$

$$u_i g_i(x) = r > 0, \quad i = 1, \ldots, m, \tag{3.2}$$

$$u_i \geq 0, \quad i = 1, \ldots, m, \tag{3.3}$$

$$\nabla f(x) - \sum_{i=1}^m u_i \nabla g_i(x) = 0, \tag{3.4}$$

and for every y such that $y^T \nabla g_i[x(r)] = 0$ (all $i \in B^*$) (under our assumptions $B^* = D^* \equiv \{i \mid u_i^* > 0\}$)

$$y^T \left[\nabla^2 f - \sum_{i \in B^*} u_i \nabla^2 g_i \right] y > 0. \tag{3.5}$$

Solving for each u_i from (3.2) and substituting in (3.4) yields

$$\nabla f[(x(r)] - \sum_{i=1}^m \frac{r}{g_i[x(r)]} \nabla g_i[x(r)] = 0. \tag{3.6}$$

Equation 3.6 says that the gradient of the function (called the *logarithmic penalty function*)

$$L(x, r) = f(x) - r \sum_{i=1}^m \ln g_i(x) \tag{3.7}$$

vanishes at $x(r)$; that is, the first-order necessary condition (2.12) for $x(r)$ to be a local unconstrained minimum of $L(x, r)$ is satisfied.

The matrix of second partial derivatives of L is

$$\nabla^2 L(x, r) = \nabla^2 f - \sum_{i=1}^m \frac{r}{g_i} \nabla^2 g_i + \sum_{i=1}^m \nabla g_i \frac{r}{g_i^2} \nabla^T g_i. \tag{3.8}$$

The subsequent results follow under suitable conditions, to be specified as the theory develops. We are tacitly assuming that as $r \to 0$, $x(r) \to x^*$. Then both $r/g_i[x(r)]$ and $r/g_i^2[x(r)] \to 0$ for all $i \notin B^*$, that is, all i, where $\lim_{r \to 0} g_i[x(r)] = g_i(x^*) > 0$. Ignoring these constraints, which contribute a negligible amount to $\nabla^2 L$ as $r \to 0$, the perturbed second-order condition (3.5) applied to (3.8) yields that $y^T \nabla^2 L[x(r), r] y > 0$ for all y such that

$$y^T \nabla g_i[x(r)] = 0 \quad \text{for all } i \in B^*.$$

(Note that this also covers the possibility that $B^* = \emptyset$.)

Since $\lim_{r \to 0} r/g_i^2[x(r)] = +\infty$ for all $i \in B^*$, it follows from (3.8) that if $B^* \neq \emptyset$, $y^T \nabla^2 L[x(r), r] y > 0$ when r is small enough, for all y such that $y^T \nabla g_i[x(r)] \neq 0$. Thus $\nabla^2 L[x(r), r]$ is a positive definite matrix satisfying the second-order sufficiency condition (2.41) that $x(r)$ be a local unconstrained minimum of $L(x, r)$.

3.1 Algorithms from Sufficiency Conditions for Constrained Minima

We have indicated that the first- and second-order sufficiency conditions for $L(x, r)$ to have an unconstrained local minimum at $x(r)$ are implied by a perturbation of the first- and second-order sufficiency conditions that hold at x^*. This was not rigorous. In particular we have not demonstrated than an $x(r)$ exists satisfying the perturbed sufficiency conditions given by (3.1–3.5). A rigorous statement of this fact under continuity and compactness assumptions is given in Section 3.3 for a general class of unconstrained functions.

A different unconstrained function results if the perturbation is modified by letting $u_i = \lambda_i^2$, $i = 1, \ldots, m$, thus automatically satisfying the non-negativity requirement (3.3) on the u_i's. Clearly $u_i g_i(x) = 0$ is equivalent to $\lambda_i g_i(x) = 0$. A feasible perturbation,

$$\lambda_i g_i = r > 0, \qquad i = 1, \ldots, m,$$

and the solution for $\lambda_i = r/g_i$, substituted into (3.4), yields

$$\nabla f[x(r)] - \sum_{i=1}^{m} \frac{r^2}{g_i^2[x(r)]} \nabla g_i[x(r)] = 0. \tag{3.9}$$

This means that the gradient of

$$P(x, r) = f(x) + r^2 \sum_{i=1}^{m} \frac{1}{g_i(x)} \tag{3.10}$$

vanishes at $x(r)$. An analysis for this P function analogous to that undertaken for the logarithmic penalty function $L(x, r)$ indicates that the first- and second-order conditions for x^* to be a local constrained minimum imply that $x(r)$ (for r small) satisfies the first- and second-order sufficiency conditions for $x(r)$ to be a local unconstrained minimum of $P(x, r)$.

The unconstrained minimization algorithms implied by these comments can be summarized as follows. Find an unconstrained local minimum of either $L(x, r)$ or $P(x, r)$ in the region where the constraints of Problem B are strictly satisfied. When r is very small an *unconstrained* local minimum of either of these functions should be close to a local constrained minimum.

Unconstrained minimization techniques of the class represented by the P and L functions are called "interior point" methods. They proceed through the interior of the region towards a solution. Another class of unconstrained minimization techniques can be described as "exterior point"; that is, they proceed toward a local minimum from the infeasible region. A development of the latter techniques is contained in Chapter 4. In the remainder of this chapter we describe in detail the intuitive and theoretical justification for interior point techniques.

3.2 GENERAL STATEMENT OF INTERIOR POINT MINIMIZATION ALGORITHMS AND THEIR INTUITIVE BASIS

A general statement of the class of "interior point" unconstrained minimization algorithms that apply to the inequality-constrained problem,

$$\text{minimize } f(x) \tag{B}$$

subject to

$$g_i(x) \geq 0, i = 1, \ldots, m,$$

is given as follows.

Let I be a scalar-valued function of x with the following two properties: Property 1 is that $I(x)$ is continuous in the region

$$R^0 = \{x \mid g_i(x) > 0, i = 1, \ldots, m\};$$

Property 2 states that, if $\{x^k\}$ is any infinite sequence of points in R^0 converging to x_B such that $g_i(x_B) = 0$ for at least one i, then $\lim_{k \to \infty} I(x^k) = +\infty$.

Let $s(r)$ be a scalar-valued function of the single variable r with the following properties. If $r_1 > r_2 > 0$, then $s(r_1) > s(r_2) > 0$. If $\{r_k\}$ is an infinite sequence of points such that $\lim_{k \to \infty} r_k = 0$, then $\lim_{k \to \infty} s(r_k) = 0$.

Definition. *An interior point unconstrained minimization technique is one that proceeds as follows.*

1. Define the function $U(x, r_1) \equiv f(x) + s(r_1)I(x)$, where r_1 is a positive number. As a starting point determine an $x^0 \in R^0$. If such a point is not readily available it may be obtained by repeated application of the method itself (see Section 8.5).

2. Proceed from x^0 to a point $x(r_1)$ that is a local minimum of U in the feasible region $R = \{x \mid g_i(x) \geq 0, i = 1, \ldots, m\}$. Presumably $x(r_1)$ will be *unconstrained* in that it will lie in R^0; otherwise $U = +\infty$, contradicting the supposition that $x(r_1)$ was a local unconstrained minimum of U in R.

3. Starting from $x(r_1)$ find a local minimum of $U(x, r_2)$, where $r_1 > r_2 > 0$.

4. Continuing in this fashion, find a local minimum for $U(x, r_k)$ starting from $x(r_{k-1})$ for a strictly monotonically decreasing sequence $\{r_k\}$.

The conjecture (to be proved) is that under appropriate assumptions the sequence of local unconstrained minima exists and has limit points that are local solutions of Problem B.

Intuitive Basis

The term $s(r)I(x)$ may be regarded as a penalty term added to the objective function $f(x)$, ensuring that a minimum of the U function is achieved in the

3.2 Interior Point Minimization Algorithms and Their Intuitive Basis

interior of the feasible region by balancing the avoidance of the boundaries and the minimization of $f(x)$. This can be seen intuitively. Consider a trajectory of steepest descent (greatest rate of decrease) of $U(x, r_1)$ starting from x^0. By assumption, $g_i(x^0) > 0$ for all i, and so $U(x, r_1)$ exists and has some finite value. Since the stipulated trajectory defines a curve along which U is continuously decreasing, no point on the trajectory can yield a U value exceeding $U(x^0, r_1)$. Since the feasible boundary is defined by $g_i(x) = 0$ (for at least one i), $U \to +\infty$ as the boundary is approached from any interior point. Consequently the boundary can never be reached by the trajectory described and the (assumed existing) minimum of $U(x, r_1)$ must be a feasible (interior) point.

In general, then, the minimum of $U(x, r)$ is feasible and does not lie on the boundary for any $r > 0$. Furthermore, when r is reduced as prescribed by the method, the weighting of the penalty factor is decreased, while the weighting of the objective function is increased. Consequently further progress can be made in minimizing $f(x)$, always maintaining feasibility.

One feature motivating this approach has already been suggested: the objective function can be reduced in value, simultaneously assuring nonviolation of the constraints. Intuitively, adequate sustainment of these properties should lead to a local feasible minimum.

Another motivation for the transformation of the original constrained problem into a sequence of unconstrained problems is that a number of methods for minimizing an unconstrained function are known and many more are being developed. Thus, if the transformation is valid, it becomes possible to solve the more formidable constrained problem by utilizing these procedures without recourse to new techniques.

Another highly desirable feature of the transformation approach is that it avoids the necessity of coping separately with the boundary of the feasible domain, for instance, attempting to move along the boundary once it is encountered. Such motion is exceedingly cumbersome when the constraining surface is nonlinear. The U function couples the objective function and the constraints in a manner that makes it possible to eliminate motion along the boundary completely.

To illustrate how an interior point unconstrained algorithm works we consider the following example in some detail.

Example.

$$\text{minimize } x_1 + x_2$$

subject to

$$g_1 = -x_1^2 + x_2 \geq 0,$$

$$g_2 = x_1 \geq 0.$$

For the U function we use the logarithmic penalty function derived in Section 3.1. Thus $s(r) = r$, and $I(x) = -\sum_{i=1}^{m} \ln g_i(x)$. The choice of s and I satisfies the requirements stated before. Then $L(x, r) = x_1 + x_2 - r \ln (-x_1^2 + x_2) - r \ln x_1$. This simple problem can be solved analytically using the fact that it is twice differentiable. Using the first-order necessary condition (2.12),

$$1 + \frac{r(2x_1)}{-x_1^2 + x_2} - \frac{r}{x_1} = 0$$

and

$$1 - \frac{r}{-x_1^2 + x_2} = 0.$$

Solving yields

$$x_1(r) = \frac{-1 \pm \sqrt{1 + 8r}}{4}.$$

Since $x_1(r)$ must be positive, only the root $x_1(r) = (-1 + \sqrt{1 + 8r})/4$ is of interest. Then $x_2(r)$ becomes

$$x_2(r) = \frac{(-1 + \sqrt{1 + 8r})^2}{16} + r.$$

That these values for $x_1(r)$ and $x_2(r)$ are local minima readily follows by observing that they satisfy the sufficient conditions (2.39) and (2.41) of Section 2.3.

Table 1 Values of r, $x(r)$ for Example

	r	$x_1(r)$	$x_2(r)$
r_1	1.000	0.500	1.250
r_2	0.500	0.309	0.595
r_3	0.250	0.183	0.283
r_4	0.100	0.085	0.107

In Table 1 are shown the computed values of $x(r)$ for four different values of r. In Figure 3 the problem is shown geometrically and indicates the points corresponding to these values of r. In the limit, the minimizing points approach the solution $(0, 0)^T$ as $r_k \to 0$.

In this problem there is only *one* unconstrained local minimum for each value of r. The problem happens to have a unique solution. It turns out that in problems with many local minima there is (subject to a mild regularity condition) a sequence of local unconstrained minima converging to each set of

3.3 Convergence Proofs for Interior Point Algorithms

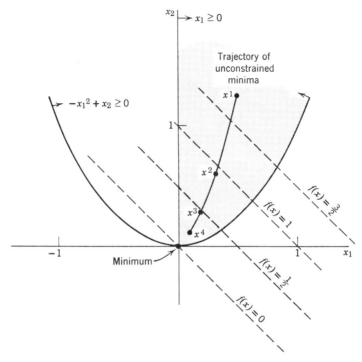

Figure 3 Solution of example by the unconstrained logarithmic function. The shaded area is the feasible region. The problem is:

$$\text{minimize } f(x_1, x_2) \equiv x_1 + x_2$$

subject to

$$g_1(x_1, x_2) \equiv -x_1^2 + x_2 \geq 0$$
$$g_2(x_1, x_2) \equiv x_1 \geq 0.$$

constrained local minima. The precise statement and proof of this fact are given in the next section.

3.3 CONVERGENCE PROOFS FOR INTERIOR POINT ALGORITHMS

In this section the existence of unconstrained U-function minima converging to local constrained minima of Problem B, assuming certain topological properties, is proved. The definitions of the usual terms of analysis are assumed known to the reader and can be found in such texts as [102] and [3]. The following definitions and lemmas are intended to make more precise the concepts of "local" minima, constrained and unconstrained.

We state without proof the following well-known result (see [102], page 67).

Lemma 2. If f is a continuous function of x in a nonempty compact set V, then there exists a finite value v^* and a point $x^* \in V$, where

$$f(x^*) = v^* = \min_V f(x).$$

For our purposes the following corollary is useful.

Corollary 8. Suppose R is a closed set, S is a compact set, and $R^0 \cap S \neq \phi$ where R^0 denotes the interior of R. If $F(x)$ is a continuous function on $R^0 \cap S$ with the property that for every sequence $\{x^k\}$ with $x^k \in R^0 \cap S$ and $x^k \to y \in (R - R^0) \cap S$, $\lim_{k \to \infty} F(x^k) = +\infty$; then there exists a finite value $\bar{v}$ and a point $\bar{x} \in R^0 \cap S$ such that $F(\bar{x}) = \bar{v} = \min_{R^0 \cap S} F(x)$.

Proof. Let x^0 be some point in $R^0 \cap S$ and define $W \equiv \{x \mid F(x) \leq F(x^0), x \in R^0 \cap S\}$. If $\{x^k\}$ is a sequence where each $x^k \in W$, then if $x^k \to y$, $y \in R \cap S$ since $R \cap S$ is compact. Hence $y \in R^0 \cap S$ or $y \in (R - R^0) \cap S$. However in the latter case $F(x^k) \to +\infty$ so that $x^k \notin W$ for k large, a contradiction. Therefore $y \in R^0 \cap S$. Further, since $F(x^k) \leq F(x^0)$, all k, the continuity of F in $R^0 \cap S$ implies that $F(y) \leq F(x^0)$. Thus W is closed as well as bounded, i.e. compact. Thus $F(x)$ attains its minimum in W. The conclusion follows from the fact that $\inf_{R^0 \cap S} F(x) = \min_W F(x)$. Q.E.D.

These theorems will be used to prove the existence of points that are local minima and local unconstrained minima, which are defined in the following way, consistent with conventional definitions.

Definition. A point $x^* \in R$ is a finite local minimum *of Problem B if* x^* *is in the* interior *of a compact set V for which*

$$f(x^*) = v^* = \min_{R \cap V} f(x).$$

Definition. Let $U(x, r)$ be an interior unconstrained minimization function. Then a point $x(r)$ is a finite local unconstrained minimum *of* $U(x, r)$ *if there exists a compact set V such that*

$$U[x(r), r] = \min_{V \cap R^0} U(x, r)$$

and $x(r)$ is contained in the interior of V.

Definition. A nonempty set $M^* \subset M$ is called an isolated set of M if there exists a closed set E such that $E^0 \supset M^*$ and such that if $x \in E - M^*$ then $x \notin M$.

Before proving the convergence of the algorithms, a fundamental theorem regarding compact sets of local minima is needed. This theorem will also be

3.3 Convergence Proofs for Interior Point Algorithms 47

used later for the class of "exterior point" unconstrained minimization techniques (see Section 4.2). It says in effect that if a set of local minima is compact it is contained in the interior of another compact set such that the local minima are global minima in the latter set.

Theorem 7 [Existence of Compact Perturbation Set]. If a set of local minima A^* corresponding to the local minimum value v^* of Problem B is a nonempty isolated compact set of $A = \{x \in R | f(x) = v^*\}$, then there exists a compact set S such that $A^* \subset S^0$, and for any point $y \in R \cap S$, if $y \notin A^*$, then $f(y) > v^*$.

Proof. Since $A^* \neq \emptyset$, $A^* \subset A$ and A^* is an isolated set of A, there exists a (nonempty) closed set E such that $E^0 \supset A^*$ and such that if $x \in E - A^*$, then $x \notin A$. Given any $y \in A^*$, there exists a (open) neighborhood $N(y)$ of y such that $f(y) = v^* \leq f(x)$ for every $x \in N(y) \cap R$. Consider the set $G = \bigcup_{y \in A^*} N(y)$, i.e., the union of all such neighborhoods $N(y)$ over A^*. Clearly, G is open, $G \supset A^*$, and $f(x) \geq v^*$ for every $x \in G \cap R$. Let S be any compact set such that $A^* \subset S^0 \subset S \subset G \cap E$.

To complete the proof, we must show that if $y \in R \cap S$ and $y \notin A^*$, then $f(y) > v^*$. Since $y \in R \cap S \subset R \cap G$, we must have $f(y) \geq v^*$ for every $y \in R \cap S$. Also, since $y \in R \cap S$ implies that $y \in E$, it follows from our first result that $f(y) \neq v^*$ if $y \in (R \cap S) - A^*$. Hence, we must have $f(y) > v^*$ if $y \in (R \cap S) - A^*$ and the proof is complete. Q.E.D.

For computational purposes it is necessary to relate the set S more directly to the problem functions. If the latter are continuous it can be shown that we can define $S \equiv \{x \mid g_i(x) \geq -\lambda, i = 1, \ldots, m, f(x) \leq v^* + \lambda\}$ providing $\lambda > 0$ and small enough.

We are now in a position to prove the convergence of the algorithms to local solutions of the programming Problem B.

Theorem 8 [Convergence to Compact Sets of Local Minima by Interior Point Algorithms]. If (a) the functions $f, g_1, \ldots, g_m$ are continuous, (b) the function $U = f(x) + s(r)I(x)$ is an interior unconstrained minimization function (I and s satisfy Properties 1 and 2), (c) a set of points A^* that are local minima corresponding to the local minimum value v^* is a nonempty, isolated compact set, (d) at least one point in A^* is in the closure of R^0, and (e) $\{r_k\}$ is a strictly decreasing null sequence, then

(i) there exists a compact set S as given in Theorem 7 such that $A^* \subset S^0$ and for r_k small the *unconstrained minima* of U in $R^0 \cap S^0$ exist and every limit point of any subsequence $\{x^k\}$ of the minimizing points is in A^*,

(ii) $\lim_{k \to \infty} s(r_k) I[x(r_k)] = 0$,
(iii) $\lim_{k \to \infty} f[x(r_k)] = v^*$,
(iv) $\lim_{k \to \infty} U[x(r_k), r_k] = v^*$,
(v) $\{f[x(r_k)]\}$ is a monotonically decreasing sequence, and
(vi) $\{I[x(r_k)]\}$ is a monotonically increasing sequence.

Proof. The proof of Parts i–iv is a by-product of the proof of Theorem 10. Let f^k denote $f[x(r_k)]$, and so on. Then, because each x^k is a global unconstrained minimum in the interior of the compact set S about A^*,

$$f^k + s(r_k) I^k \leq f^{k+1} + s(r_k) I^{k+1},$$

$$f^{k+1} + s(r_{k+1}) I^{k+1} \leq f^k + s(r_{k+1}) I^k.$$

Adding $s(r_k)/s(r_{k+1})$ times the first inequality to the second, rearranging, and dividing [by our assumptions on $\{r_k\}$ and $s(r)$, $s(r_k) > s(r_{k+1})$] yields

$$f^k \geq f^{k+1}.$$

This proves Part v. Using this and the first inequality above proves Part vi.
Q.E.D.

The following example demonstrates an application of the theorem.

Example.
$$\text{minimize } x_2$$
subject to
$$x_2 - \sin x_1 - \frac{x_1}{2} \geq 0.$$

Consider the U function to be the logarithmic function

$$L(x, r) = x_2 - r \ln \left[x_2 - \sin x_1 - \frac{x_1}{2} \right]. \tag{3.11}$$

Since these functions are twice differentiable we use first and second derivatives to determine the local unconstrained minima of (3.11). Using (2.12),

$$\frac{r[\cos(x_1) + \tfrac{1}{2}]}{[x_2 - \sin(x_1) - x_1/2]} = 0$$

and

$$1 - \frac{r}{[x_2 - \sin(x_1) - x_1/2]} = 0.$$

3.3 Convergence Proofs for Interior Point Algorithms

There are two sets of possible solutions:

$$x_1(r) = \frac{2\pi}{3} \pm 2n\pi, \quad n = 0, 1, 2, 3, \ldots, \tag{3.12}$$

$$x_2(r) = \sin\left[\frac{2\pi}{3} \pm 2n\pi\right] + \frac{\pi}{3} \pm n\pi + r, \quad n = 0, 1, \ldots \tag{3.13}$$

and

$$x_1(r) = \frac{4\pi}{3} \pm 2n\pi, \quad n = 0, 1, 2, \ldots, \tag{3.14}$$

$$x_2(r) = \sin\left(\frac{4\pi}{3} \pm 2n\pi\right) + \frac{2\pi}{3} \pm n\pi + r, \quad n = 0, 1, \ldots. \tag{3.15}$$

The matrix of the second partial derivatives of (3.11) is

$$\begin{bmatrix} -\sin[x_1(r)] & 0 \\ 0 & \frac{1}{r} \end{bmatrix}.$$

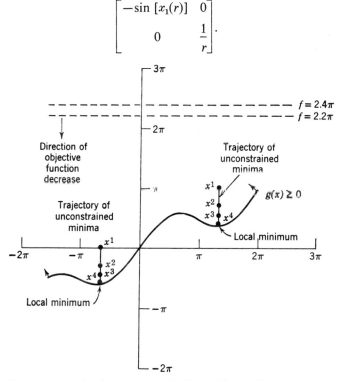

Figure 4 Convergence to local unconstrained minima. The problem is:

$$\text{minimize } x_2$$

subject to

$$x_2 - \sin x_1 - \frac{x_1}{2} \geq 0.$$

Table 2 Values of Two Local Minima for Example

		$(n = 0)$		$(n = -1)$	
	r	$x_1(r)$	$x_2(r)$	$x_1(r)$	$x_2(r)$
r_1	2.0	$\frac{4}{3}\pi$	$(1.027)\pi$	$-\frac{2}{3}\pi$	0.028π
r_2	1.0	$\frac{4}{3}\pi$	$(0.709)\pi$	$-\frac{2}{3}\pi$	-0.290π
r_3	0.5	$\frac{4}{3}\pi$	$(0.550)\pi$	$-\frac{2}{3}\pi$	-0.449π
r_4	0.1	$\frac{4}{3}\pi$	$(0.423)\pi$	$-\frac{2}{3}\pi$	-0.576π

For $x_1(r) = \frac{2}{3}\pi \pm 2n\pi$, this is *not* a positive semidefinite matrix. For $x_1(r) = \frac{4}{3}\pi \pm 2n\pi$, the matrix is positive definite and therefore satisfies the sufficient conditions for a local minimum. Thus there are an infinite number of L-function trajectories, one for each local minimum of the problem in this example. Two of these are shown in Figure 4. The data are in Table 2.

Several corollaries follow from Theorem 8. They are stated without proof.

Corollary 9. If in addition to the Assumptions of Theorem 8, A^* is the set of global minima and A^* is bounded, then the conclusions of Theorem 8 apply and the $\{x(r_k)\}$ are global unconstrained minima of $U(x, r_k)$ in R^0. (The proof follows because bounded sets of *global* minima must be *closed* also, hence compact. This is not true of *local* minima.) (See Stong [104] for proof of a similar theorem for a general topological space.)

Corollary 10. If x^* is an isolated local minimum corresponding to the local minimum value v^*, and if Assumptions a, b, d, and e of Theorem 8 are satisfied, then the sequence $\{x(r_k)\}$ of that theorem is such that $\{x(r_k)\}$ converges to x^*.

Regularized Interior Point Methods

By incorporating a suitable regularization in the interior point auxiliary function associated with Problem B, Assumption d regarding the specified closure property of R^0 can be eliminated. In particular, it is possible to slack the constraints by a factor that depends on the parameter r and vanishes as $r \to 0$. Defining $U(x, r)$ as in the above definition, in the relaxed constraint region, then leads to the desired minimizing sequence (in the enlarged domain) as in Conclusion i, and yields Conclusions ii, iii, and iv as well.

As an example, a regularization of the interior point function (3.10) given

3.3 Convergence Proofs for Interior Point Algorithms

in Section 3.1 is given by

$$P(x, r) = f(x) + r^2 \sum_{i=1}^{m} \frac{1}{g_i(x) + r}.$$

A proof of a generalization of the above procedure is given in [48].

The desired partial converse of Theorem 8 is not true. In particular, it is possible that a sequence of unconstrained U-function minima may converge to a saddle point of the programming problem where the necessary, but not the sufficient, conditions for a local minimum are satisfied.

Example. Consider the programming problem in E^1,

$$\text{minimize } \frac{(x_1 - 1)^3}{3}$$

subject to

$$x_1 + 1 \geq 0.$$

Let $s(r) = r^2$ and $I(x) = 1/g(x)$. Clearly there is only one local (and hence global) solution to this problem, at $x = -1$. However, there is a saddle point at $x_1 = 1$, where the necessary but not sufficient conditions for $x_1 = 1$ to be a local minimum are satisfied. The U function here is

$$U(x, r) = \frac{(x_1 - 1)^3}{3} + \frac{r^2}{x_1 + 1}. \tag{3.16}$$

Taking the first derivative and setting it equal to 0 yields

$$(x_1 - 1)^2 - \frac{r^2}{(x_1 + 1)^2} = 0. \tag{3.17}$$

For any value of r there are four possible solutions to (3.17) given by $x_1(r) = \pm\sqrt{1 \pm r}$. The root $x_1(r) = -\sqrt{1 + r}$ is ruled out since $x(r) + 1 > 0$ must be satisfied. The second partial derivative of (3.16) is

$$2(x_1 - 1) + \frac{2r^2}{(x_1 + 1)^3}.$$

For a local minimum this must be non-negative. The root $x(r) = \sqrt{1 - r}$ is a local maximum when $r < 1$. The root $x(r) = -\sqrt{1 - r}$ is a local minimum (for $r < 1$) and converges to the solution.

The last root, $x(r) = \sqrt{1 + r}$, is a local minimum of U but converges to the saddle point of the objective function. This root is the only one that exists for $r > 1$.

Any nonlinear programming method encountering $x_1 = 1$ would have a difficult time, since this point satisfies all indications that it is indeed a local solution to the problem.

We shall return to interior point methods after we discuss a class of unconstrained minimization techniques that approach the optimum from the infeasible region.

4

Exterior Point Unconstrained Minimization Techniques

4.1 GENERAL STATEMENT OF EXTERIOR POINT ALGORITHMS AND THEIR INTUITIVE DERIVATION

Like the derivation of interior point algorithms given in Chapter 3, the following provides a development for exterior point algorithms, based again on the perturbation of the sufficiency conditions for a local minimum of Problem B.

Consider a relaxation of the constraints of Problem B of the form

$$g_i(x) \geq -r, \quad i = 1, \ldots, m, \tag{4.1}$$

where r is a positive number. Assume that when r is small, near any local constrained minimum x^*, an $x(r)$ exists satisfying (4.1) and also satisfies the following conditions.

We perturb the complementarity condition (2.28) so that it makes sense for negative values of g_i and reduces to the proper equation as $r \to 0$. Essentially the complementarity condition says that if $g_i(x^*) > 0$, then $u_i^* = 0$, and if $g_i(x^*) = 0$, u_i^* can be any quantity and still satisfy the complementarity equation. We let

$$u_i r = -\min\,[0, g_i(x)] \tag{4.2}$$

be the perturbed complementarity condition. Thus if, when r is small, the $x(r)$ satisfying (4.1) for some i yields $g_i[x(r)] \geq 0$, then $u_i \equiv 0$ and $\lim_{r \to 0} u_i(r) = u_i^* = 0$. If $g_i[x(r)] < 0$, $\lim_{r \to 0} -\min\,(0, g_i[x(r)]) = 0$, since (4.1) guarantees feasibility as $r \to 0$. Since by (4.1) $g_i[x(r)] \geq -r$, and by (4.2) $\lim_{r \to 0} u_i(r) r = 0$, then $\lim_{r \to 0} u_i(r) g_i[x(r)] = 0$, and the final part of the complementarity condition is satisfied in the limit. [Note also that the requirement $u_i \geq 0$, $i = 1, \ldots, m$ is satisfied by (4.2).]

Exterior Point Unconstrained Minimization Techniques

The analogous perturbation of the remaining sufficiency conditions (2.29–2.31) for a local constrained minimum is

$$u_i(r) \geq 0, \qquad i = 1, \ldots, m, \tag{4.3}$$

$$\nabla f[x(r)] - \sum_{i=1}^{m} u_i(r) \nabla g_i[x(r)] = 0, \tag{4.4}$$

and for every vector y such that $y^T \nabla g_i[x(r)] = 0$, for all $i \in D^* = \{i \mid u_i^* > 0\}$, it follows that

$$y^T \left\{ \nabla^2 f[x(r)] - \sum_{i \in D^*} u_i(r) \nabla^2 g_i[x(r)] \right\} y > 0. \tag{4.5}$$

Solving (4.2) for each i and substituting in (4.4) yields

$$\nabla f[x(r)] + \sum_{i=1}^{m} \frac{\min \{0, g_i[x(r)]\}}{r} \nabla g_i[x(r)] = 0. \tag{4.6}$$

Equation 4.6 states that the first-order necessary condition (2.12) is satisfied that $x(r)$ be a local unconstrained minimum of the function whose gradient is (4.6). This function must be

$$f(x) + \sum_{i=1}^{m} \frac{\{\min [0, g_i(x)]\}^2}{2r}. \tag{4.7}$$

If we make the assumption that the gradients of the constraints for all $i \in B^*$ form a linearly independent set, then each $u_i(r)$, as obtained from (4.2), has as its limit the unique u_i^*, where (x^*, u^*) satisfies the hypotheses of Theorem 4.

A further assumption, that strict complementarity holds in (2.28), implies, along with (4.2), that $g_i[x(r)] < 0$ for all $i \in D^*$. It also implies that

$$g_i[x(r)] > 0$$

for all $i \notin D^*$, since strict complementarity implies $B^* = D^*$, where $B^* = \{i \mid g_i(x^*) = 0\}$.

Then the matrix of second partial derivatives of (4.7) is given by

$$\nabla^2 f(x) + \sum_{i \in D^*} \frac{\min [0, g_i(x)]}{r} \nabla^2 g_i(x) + \sum_{i \in D^*} \nabla g_i(x) \frac{1}{r} \nabla^T g_i(x). \tag{4.8}$$

[The function (4.7) is not twice differentiable when $g_i(x) = 0$. Under our assumptions this possibility is ruled out when r is small.]

Under all these assumptions it is possible in a manner identical with that in Section 3.1 to show, using (4.5), that (4.8) is a positive definite matrix and hence that $x(r)$ is a local constrained minimum of (4.7).

4.1 Exterior Point Algorithms and Their Intuitive Derivation

The preceding analysis then implies that finding an unconstrained minimum of (4.7) for values of r that approach zero will yield a sequence converging to a constrained minimum of Problem B.

The unconstrained function (4.7) is called the "quadratic loss" function and is an example of a class of unconstrained minimization techniques called "exterior point" methods. A general description of exterior point algorithms can be given as follows. Initially these will be applied to problems where only continuity of the problem functions is assumed.

Let $p(t)$ be a scalar-valued function of the single variable t with the properties that if $0 < t_1 < t_2$ then $0 < p(t_1) < p(t_2)$, and if $\{t_k\}$ is a monotonically increasing sequence of positive values where $\lim_{k \to \infty} t_k = +\infty$, then $\lim_{k \to \infty} p(t_k) = +\infty$. Let $O(x)$ be a continuous function of x with the properties that $O(x) = 0$ if $g_i(x) \geq 0$, $i = 1, \ldots, m$ (that is, if $x \in R$), and $O(x) > 0$ otherwise. Then $T(x, t) = f(x) + p(t)O(x)$ is an exterior point minimization function.

Definition. *The exterior point unconstrained minimization technique is as follows.*

1. For some $t_1 > 0$, find an unconstrained local minimum of $T(x, t_1)$. Denote it by $x(t_1)$.
2. Given $x(t_1)$, find an unconstrained local minimum of $T(x, t_2)$, where $t_2 > t_1$.
3. Proceed in this fashion, minimizing $T(x, t_k)$ for a strictly monotonically increasing (to $+\infty$) sequence $\{t_k\}$.

Presumably as $t_k \to +\infty$ the sequence of unconstrained local minima will approach a local constrained minimum x^*.

The intuitive basis for this can be seen as follows. If x strays too far from the feasible region the penalty term $p(t_k)O(x)$ becomes large when t_k is large. Thus as $t_k \to +\infty$ the tendency will be to draw the unconstrained minima toward the feasible region so as to minimize the value of the penalty term. That is, a large penalty is attached to being infeasible. This is called an exterior point algorithm because the movement is from the outside or infeasible region toward the inside of the feasible region.

We demonstrate the use of such an algorithm by solving the following example.

Example.
$$\text{minimize } -x_1 x_2$$
subject to
$$g_1 \equiv -x_1 - x_2^2 + 1 \geq 0,$$
$$g_2 \equiv x_1 + x_2 \geq 0.$$

Let $O_i(g_i) = [(g_i - |g_i|)/2]^2$, let $O(x) = \sum O_i[g_i(x)]$, and let $p(t) = t$. Then

$$T = -x_1 x_2 + t\left[\frac{(-x_1 - x_2^2 + 1 - |-x_1 - x_2^2 + 1|)}{2}\right]^2$$

$$+ t\left[\frac{(x_1 + x_2 - |x_1 + x_2|)}{2}\right]^2 \quad (4.9)$$

is an exterior point unconstrained minimization function.

In Table 3 are given the values of $x(t)$ for four different values of t. These are plotted in Figure 5 and are seen to approach the optimum from the region of infeasibility as t increases.

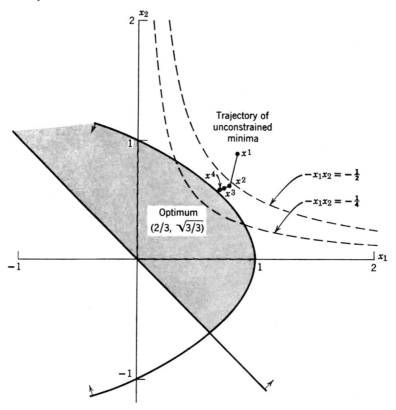

Figure 5 Solution of example by an exterior unconstrained algorithm. The shaded area is the feasible region. The problem is:

$$\text{minimize } -x_1 x_2$$

subject to

$$-x_1 - x_2^2 + 1 \geq 0,$$
$$x_1 + x_2 \geq 0.$$

Table 3 Values of $x(t)$ for Example

t		$x_1(t)$	$x_2(t)$
t_1	1.0	0.89	0.67
t_2	2.0	0.77	0.62
t_3	3.0	0.73	0.61
t_4	10.0	0.67	0.58
Optimum		2/3	$\sqrt{3}/3$

There are several important differences between interior point and exterior point algorithms other than the obvious one that the former are always feasible and the latter are usually infeasible. These differences are discussed at length in Section 4.3. In the next section a very important similarity to the interior point algorithm—the convergence to compact sets of local minima of the exterior point algorithm—is proved. The proof is based on the fundamental Theorem 7 and needs only the continuity of the problem functions.

4.2 CONVERGENCE THEOREM FOR EXTERIOR POINT ALGORITHMS

In this section a theorem on the existence of unconstrained T-function minima converging to local solutions of Problem B analogous to Theorem 8 for interior point algorithms is given. The essential difference in the requirements on the problem is that in this instance the interior of the feasible region (R^0) need not be nonempty.

Theorem 9 [Convergence to Compact Sets of Local Minima by Exterior Point Algorithms]. If (a) the functions $f, g_i, \ldots, g_m$ are continuous, (b) the function $T = f(x) + p(t)O(x)$ is an exterior point unconstrained minimization function (that is, p and O satisfy the properties stated in Section 4.1), (c) a set of points A^*, which are local minima with minimum value v^* of Problem B, is a nonempty, isolated compact set, and (e) $\{t_k\}$ is a strictly increasing unbounded non-negative sequence then

(i) there exists a compact set S as given in Theorem 7 such that $A^* \subset S^0$ and for t_k large enough the unconstrained minima of $T(x, t_k)$ in S^0 exist and every limit point of any subsequence $\{x^k\}$ of the minimizing points is in A^*;
(ii) $\lim_{k \to \infty} p(t_k)O[x(t_k)] = 0$;
(iii) $\lim_{k \to \infty} f[x(t_k)] = v^*$;
(iv) $\lim_{k \to \infty} T[x(t_k), t_k] = v^*$, $T[x(t_k), t_k] \leq v^*$, all k;
(v) $\{f[x(t_k)]\}$ is a monotonically increasing sequence; and
(vi) $\{O[x(t_k)]\}$ is a monotonically decreasing sequence.

58   Exterior Point Unconstrained Minimization Techniques

Proof. Parts i–iv are a by-product of Theorem 10. Let f^k denote $f[x(t_k)]$, and so on. Then, because each x^k is a global unconstrained minimum in the interior of a compact set containing A^*,

$$f^k + p(t_k)O^k \le f^{k+1} + p(t_k)O^{k+1}$$

and

$$f^{k+1} + p(t_{k+1})O^{k+1} \le f^k + p(t_{k+1})O^k.$$

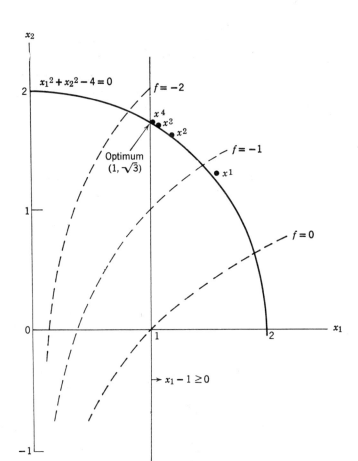

Figure 6 Solution of example by a mixed algorithm. The problem is:

$$\text{minimize } \ln x_1 - x_2$$

subject to

$$x_1 - 1 \ge 0,$$
$$x_1^2 + x_2^2 - 4 = 0.$$

Adding $p(t_{k+1})/p(t_k)$ times the first inequality to the second, and canceling, yields

$$\left[\frac{p(t_{k+1})}{p(t_k)} - 1\right] f^k \leq \left[\frac{p(t_{k+1})}{p(t_k)} - 1\right] f^{k+1}.$$

Using our assumptions on $\{t_k\}$ and $p(t)$, Part v follows. Using (v) and the second inequality above proves (vi).   Q.E.D.

As with the interior point algorithm, several corollaries follow from this theorem.

Corollary 11. If a set of points that are global minima of f in R constitute a *bounded* isolated set, then the conclusions of Theorem 9 apply for this set.

It is interesting that unlike the corresponding result for the interior point method, the minimizing sequence $\{x^k\}$ need not be global minima.

Corollary 12. If x^* is an isolated local minimum with local minimum value v^*, then the sequence $\{x(t_k)\}$ of Theorem 9 is such that $x(t_k)$ converges to x^*.

Note that interiority is not needed. Thus the results derived here are even more general than the results of Theorem 8.

Note also that equality constraints can be handled by simply rewriting each equality as two inequalities, in the obvious manner, assuming only that the resulting system subscribes to the conditions required for Theorem 9. (This handling of equalities and the resulting form of the unconstrained function is indicated in the example of Figure 6.)

4.3 MIXED INTERIOR POINT–EXTERIOR POINT ALGORITHMS

For certain problems, for example, problems with equality constraints, interior methods do not apply since there is no "interior" to the region of feasibility. For such problems we would like to handle that subset of constraints differently. For other reasons, for example, when continued satisfaction of certain constraints is demanded, it is useful to develop a mixed algorithm, one that maintains strict satisfaction of some of the constraints as computations proceed, requiring others to be satisfied only as the solution is approached. In the following discussion a straightforward mixed algorithm is developed, combining the methods developed above and the theoretical convergence proved under the same conditions of compactness and continuity as for the two "pure" methods.

We assume that the programming problem is written in the following form.

$$\text{minimize } f(x) \qquad \text{(M)}$$

subject to

$$g_i(x) \geq 0, \qquad i = 1, \ldots, m, \qquad (4.10)$$

$$g_j(x) \geq 0, \qquad j = m+1, \ldots, q. \qquad (4.11)$$

We define $Q \equiv \{x \mid g_j(x) \geq 0, j = m+1, \ldots, q\}$, and R and R^0 will be as before. With respect to the constraints (4.10) we require that strict satisfaction hold at all points of our algorithm. The constraints (4.11) will be handled by an exterior point penalty function and their satisfaction will be required only as the penalty $p(t_k)$ goes to $+\infty$.

Definition. We define our mixed unconstrained minimization function as

$$V(x, r_k, t_k) = f(x) + s(r_k)I(x) + p(t_k)O(x). \qquad (4.12)$$

We assume that (s, I) and (p, O) have the same properties for V that they had for U and T, respectively. The only difference is that I is defined for x strictly satisfying (4.10), and O is defined relative to (4.11).

Theorem 10 [Convergence of Mixed Algorithm]. If (a) the functions f, $g_1, \ldots, g_q$ are continuous, (b) $s(r)$, $I(x)$, $p(t)$, and $O(x)$ satisfy the properties defining interior and exterior penalty functions, (c) a set of points A^*, which are local minima of Problem M corresponding to the local minimum value v^*, is a nonempty isolated compact set, (d) at least one point in A^* is in the closure of $R^0 \cap Q$, and (e) $\{r_k\}$ is a strictly decreasing null sequence and $\{t_k\}$ is a non-negative, strictly increasing unbounded sequence, then

(i) there exists a compact set S, given in Theorem 7, such that $A^* \subset S^0$, and for k large enough the unconstrained minima of $V(x, r_k, t_k)$ in $R^0 \cap S^0$ exist and every limit point of any subsequence $\{x^k\}$ of the minimizing points is in A^*;
(ii) $\lim_{k \to \infty} s(r_k)I(x^k) = 0$;
(iii) $\lim_{k \to \infty} p(t_k)O(x^k) = 0$; and
(iv) $\lim_{k \to \infty} V(x^k, r_k, t_k) = v^*$.

Proof. Because of Assumption c, Theorem 7 applies and there is a compact set S where $A^* \subset S^0$ such that $f(y) > v^*$ for $y \in R \cap Q \cap S$ and $y \notin A^*$. We shall prove that for this set S Conclusions i–iv hold.

Define x^k so that

$$V(x^k, r_k, t_k) = \min_{R^0 \cap S} V(x, r_k, t_k).$$

The existence of x^k is guaranteed by the continuity of the problem functions, the continuity of the penalty terms $I(x)$ and $O(x)$, which imply the continuity of the V function in $R^0 \cap S$, and Corollary 8. From Corollary 8 we also conclude that each $x^k \in R^0 \cap S$.

4.3 Mixed Interior Point–Exterior Point Algorithms

Let y^0 denote one of the limit points of the uniformly bounded sequence $\{x^k\}$. Clearly $y^0 \in R \cap S$. Denote the subsequence converging to y^0 also by $\{x^k\}$. We need to show that $y^0 \in A^*$. First we show that $y^0 \in Q$. If this were not so, then $\liminf_{k \to \infty} V(x^k, r_k, t_k) = +\infty$ using the fact that $p(t_k)O(x^k) \to +\infty$ (for $y^0 \notin Q$), and $f(x^k), r_k I(x^k)$ are bounded below in S. But by Assumption d there exists at least one $x^0 \in R^0 \cap Q \cap S$. For this point

$$\lim_{k \to \infty} V(x^0, r_k, t_k) = f(x^0) < +\infty.$$

This contradicts the assumption that x^k minimizes V for k large. Thus $y^0 \in R \cap Q \cap S$.

If $y^0 \notin A^*$, then by the basic property of the S set, $f(y^0) > v^*$. By Assumption d there is an $x^0 \in R^0 \cap Q \cap S$, where

$$v^* < f(x^0) < f(y^0).$$

Then

$$\liminf_{k \to \infty} V(x^k, r_k, t_k) \geq f(y^0) > f(x^0) = \lim_{k \to \infty} V(x^0, r_k, t_k). \quad (4.13)$$

This contradicts the assumption that x^k minimizes V for large k. Hence $y^0 \in A^*$. But since $A^* \subset S^0$, for k large enough x^k must be in $R^0 \cap S^0$ and hence is an unconstrained minimum. This proves Part i. Parts ii–iv follow from the fact that $\lim_{k \to \infty} f(x^k) = v^*$, and arguments used in (4.13).

Q.E.D.

Corollary 13. If, in addition to the hypotheses of Theorem 10, a set of global minima of f in $R \cap Q$ is a *bounded*, isolated set, the conclusions of Theorem 10 apply for this set.

Corollary 14. If x^* is an isolated local minimum, and if the other hypotheses of Theorem 10 are satisfied, then the sequence $\{x^k\}$ of Theorem 10 is such that x^k converges to x^*.

Example. Suppose that the programming problem to be solved is

$$\text{minimize } \ln x_1 - x_2$$

subject to

$$g_1 = x_1 - 1 \geq 0,$$
$$h_1 = x_1^2 + x_2^2 - 4 = 0.$$

If we write the equality as two inequality constraints,

$$g_2 = x_1^2 + x_2^2 - 4 \geq 0$$

and

$$g_3 = -x_1^2 - x_2^2 + 4 \geq 0,$$

we have a problem of the form (B). An interior point function cannot be applied. Suppose that we apply the quadratic loss function to the last two inequality constraints. This adds terms

$$t_k[(g_2 - |g_2|)/2]^2 + t_k[(-g_2 - |-g_2|)/2]^2$$

to whatever unconstrained function we generate. But this is just $t_k(g_2)^2$, regardless of the sign of g_2. In general, then, with respect to equality constraints, the quadratic loss function applied to the two separate inequalities $h_j \geq 0$, $-h_j \geq 0$ yields a single term $t_k h_j^2$. We assume for this example that the two terms have been combined.

If we apply an exterior function to $(x_1 - 1)$, the first inequality constraint, we cannot permit $x_1 < 0$ since the objective function is not defined there. We have here an example where strict satisfaction of a constraint is mandatory, and hence we must use a technique such as the interior point penalty function for that constraint. We choose the logarithmic penalty function. We also choose $t = r^{-1}$, and let $r \to 0^+$.

Our mixed unconstrained function is now

$$V(x, r, r^{-1}) = \ln x_1 - x_2 - r \ln (x_1 - 1) + r^{-1}[x_1^2 + x_2^2 - 4]^2.$$

Differentiating with respect to x_1 and x_2 yields

$$\frac{1}{x_1} - \frac{r}{x_1 - 1} + 4r^{-1}x_1(x_1^2 + x_2^2 - 4) = 0,$$

$$-1 + 4r^{-1}x_2(x_1^2 + x_2^2 - 4) = 0.$$

Solving these numerically yields the values given in Table 4. The plot of the minimizing sequence and the convergence to the optimum at $(1, \sqrt{3})$ are seen in Figure 6.

Table 4 Values of $x(r)$, $V[x(r), r]$ for Example

	r	$x_1(r)$	$x_2(r)$	$V[x(r), r]$
r_1	1.0	1.553	1.334	−0.2648
r_2	$\frac{1}{4}$	1.159	1.641	−1.0285
r_3	$\frac{1}{16}$	1.040	1.711	−1.4693
r_4	$\frac{1}{64}$	1.010	1.727	−1.6447
r_5	$\frac{1}{256}$	1.002	1.731	−1.7048

The convergence of the mixed interior point–exterior point algorithm can be used to prove Corollary 2, Section 2.1. For simplicity we assume that $I(g) = -\sum_{i=1}^{m} \ln g_i$, and that the quadratic loss function is applied to the equality constraints $h_j(x) = 0, j = 1, \ldots, p$. We assume that x^* is a local

4.3 Mixed Interior Point–Exterior Point Algorithms

minimum of Problem A. Further, as in Corollary 2, we assume that there exists an s having the properties stated there. We modify the objective function f to make it $f(x) + \|x - x^*\|^2$ and minimize the mixed interior-exterior point penalty function

$$\|x - x^*\|^2 + f(x) - r_k \sum_{i=1}^{m} \ln g_i(x) + r_k^{-1} \sum_{j=1}^{p} [h_j(x)]^2 \qquad (4.14)$$

for a decreasing null sequence $\{r_k\}$.

By Corollary 14, for r_k small enough there exists a sequence $\{x^k\}$ of unconstrained local minima, where x^* is the unique limit point of that sequence. (The existence of s implies that $R^0 \cap Q$ is not empty in a neighborhood of x^*.)

Using the first-order necessary condition that x^k minimize (4.14), we have

$$2(x^k - x^*) + \nabla f^k - \sum_{i=1}^{m} u_i^k \nabla g_i^k + \sum_{j=1}^{p} w_j^k \nabla h_j^k = 0. \qquad (4.15)$$

where $u_i^k = r_k/g_i^k$, $i = 1, \ldots, m$ (note $u_i^k > 0$, all k), and

$$w_j^k = \frac{2h_j^k}{r_k}, \qquad j = 1, \ldots, p.$$

Clearly $\lim_{k \to \infty} u_i^k = 0$ for all $i \notin B^*$. The limits of the other u_i^k's and w_j^k's may or may not be finite. We shall show that the existence of the s implies that the limits of (u^k, w^k) are all finite. For assume the contrary, and let

$$v^k = \sum_{i=1}^{m} u_i^k + \sum_{j=1}^{p} |w_j^k|.$$

Then we divide both sides of (4.15) by v^k and take the limit as $k \to \infty$. Clearly for some nonzero set of values $\alpha_i \geq 0$, $i \in B^*$, β_j, $j = 1, \ldots, p$,

$$-\sum_{i \in B^*} \alpha_i \nabla g_i^* + \sum_{j=1}^{p} \beta_j \nabla h_j^* = 0. \qquad (4.16)$$

If all the α_i's are equal to zero, at least one β_j is not, and the independence assumption in $\{\nabla h_j^*\}$ is violated. If some $\alpha_i > 0$, multiplying both sides of (4.16) by the given s yields $0 < 0$, a contradiction. Hence (u^k, w^k) has only finite limits. Let (u^*, w^*) be one of them, and then taking the limit as $k \to \infty$ in (4.15) yields (2.9). Q.E.D.

Differences between Interior and Exterior Point Algorithms

Interior point and exterior point methods represent essentially different philosophies. Whereas interior point methods have a penalty term that prevents constraint violation, exterior methods use the penalty term to prevent the points from straying too far from the feasible region. The latter methods are more "relaxed" in that sense.

Exterior point methods handle the combinatorial difficulties associated with inequality constraints in that presumably many constraints that are nonbinding at the solution will be naturally excluded from the computations since somewhere, in the course of the iterations, they will become feasible and remain so. Thus their first and second partial derivatives, which are usually needed (Chapter 8) to compute the unconstrained minimum of any unconstrained penalty function, need not be computed when the constraints are strictly satisfied. Interior methods, on the other hand, require information for all the constraints all the time and do not try to guess which constraints are not constraining at the solution. (Modifications to interior point algorithms that attempt to get around these difficulties are contained in Section 7.1.)

The price that exterior point algorithms pay for this advantage is reflected in the lack of orders of differentiability (in x) of the exterior point unconstrained function at any boundary point of the feasible region. The quadratic loss function, for example, is first differentiable but not second differentiable at the boundary of the feasible region. Interior methods, on the other hand, have an order of differentiability as high as that of the problem functions themselves [assuming that the function $I(g)$ has infinite order of differentiability, such as the P function, or the logarithmic penalty function]. Thus theorems for proving convergence of methods for minimizing unconstrained functions must take this into account. Furthermore, the computational implications of this are apparent when one considers that an exterior point method starting at an interior point has only first- and second-derivative information for the objective function. What sort of efficient moves should be made in this instance if the objective function is mixed linear-nonlinear? Partial answers to these questions are found in Section 8.2.

Exterior point methods do not require an interior to the feasible inequality-constrained region. Thus equality as well as inequality constraints can be handled with the same ease. Obviously interior methods are unable to handle equality constraints, and mixed methods are unavoidable unless all the equality constraints are linear. In this case mixed simplicial-unconstrained methods can be used (see Section 7.7), or regularized interior methods [48].

No separate "feasibility" phase is required for exterior point methods as is required for interior methods. This makes coding exterior algorithms a simpler matter. (An approach for gaining interior feasibility for interior methods is contained in Section 8.5.)

Finally, there is reason to believe that minimizing an unconstrained exterior point function will be harder than minimizing an interior point unconstrained function. This is based on numerical analysis considerations of the condition numbers of the Hessian matrices of the unconstrained functions [90].

All of these differences are essentially computational ones since, except for the interiority assumption on the constraint region, as seen in the last two chapters and in the next one, convergence of the two algorithms derives from the same conditions. The situations in which interior or exterior methods are preferred are not yet determined. Some comments in Chapter 8 are directed toward this topic. Certainly the complete computational questions implied by these methods have yet to be answered.

4.4 GENERALIZED INTERIOR AND EXTERIOR POINT METHOD†

Close scrutiny of the proof of Theorem 10 leads to the observation that only the continuity and a few simple limiting properties of the V function are actually utilized. It is thus possible to single out several key properties that are sufficient to guarantee the conclusion of the theorem. In so doing, we are led to a generalization of the $V(x, r, t)$ penalty function and the method it defines.

Having done this, we provide an analogous generalization of the $U(x, r)$ and $T(x, t)$ functions defined in the previous sections, and subsequently develop a whole hierarchy of penalty functions. The correspondingly generalized methods are immediate consequences. Thus the U and T functions treated in detail in the previous sections and throughout this book may be viewed as effective realizations of a very large family of penalty functions.

With Problem M as stated above, we assume $R^0 \neq \emptyset$ and $Q \neq \emptyset$. As throughout, we always tacitly assume that $x \in E^n$, and the functions $f(x)$ and $g_1(x), \ldots, g_q(x)$ are continuous.

DEFINING PROPERTIES OF $V(x, r, t)$

(i) $V(x, r, t)$ is continuous for $x \in R^0$ for any $r > 0$ and $t > 0$.
(ii) If $\{x^k\} \subset R^0$ and $x^k \to y \in R - R^0$, where $\|y\| < \infty$, then $\lim_{k \to \infty} V(x^k, r, t) = +\infty$ for any $r > 0$, $t > 0$.
(iii) If $\{x^k\} \subset R^0$, $r_k > 0$, and $t_k > 0$ for every k, and $(x^k, r_k, t_k) \to (y, 0, +\infty)$, with $\|y\| < \infty$, then $\liminf_{k \to \infty} V(x^k, r_k, t_k) \begin{cases} = +\infty \text{ if } y \notin Q, \\ \geq f(y) \text{ otherwise.} \end{cases}$
(iv) If $y \in R^0 \cap Q$, $r_k > 0$ and $t_k > 0$ for every k, and $(r_k, t_k) \to (0, +\infty)$, then $\lim_{k \to \infty} V(y, r_k, t_k) = f(y)$.

† The material in this and the following two sections is taken from [48].

With $V(x, r, t)$ so defined, the indicated generalization of Theorem 10 is valid. We state this.

Theorem 11 [Convergence of V Minima to Local Solutions of Problem M]. If $f, g_1, \ldots, g_q$ are continuous, $V(x, r, t)$ is as above, $A^* \cap \overline{R^0} \cap Q \neq \emptyset$, $A^* \equiv \{x \mid f(x) = v^*, x \text{ is a local solution of } (M)\} \neq \emptyset$ isolated and compact, $r_k > 0$, $t_k > 0$ for every k, and $(r_k, t_k) \to (0, +\infty)$, then for all k large enough there exist unconstrained local minima x^k of $V(x, r_k, t_k)$ in R^0 such that, if x^* is any limit point of the uniformly bounded sequence $\{x^k\}$, $x^* \in A^*$.

Proof. The proof parallels that of Theorem 10 and is not repeated here.

Before we proceed with the general development, a few interesting facts may be noted about the general penalty function that has been defined. Suppose that this function has the form $V[f(x), g_i(x), \ldots, g_q(x), r, t]$, V is once differentiable in $f, g_1, \ldots, g_q$, and these latter functions are differentiable in x, for $x \in R^0$ and any $r > 0$, $t > 0$. Then, if x^k is a local unconstrained minimum of V, it follows that

$$\nabla V^k = \frac{\partial V^k}{\partial f} \nabla f(x^k) + \sum_{i=1}^{q} \frac{\partial V^k}{\partial g_i} \nabla g_i(x^k) = 0.$$

This equation is the same as $\nabla \mathcal{L}(x^k, u^k) = 0$, where $\mathcal{L}$ is the Lagrangian associated with Problem M, provided that we assume $\partial V / \partial f \neq 0$ and set

$$u_i^k = \frac{-\partial V^k / \partial g_i}{\partial V^k / \partial f}, \quad i = 1, \ldots, q.$$

This shows how we can be led to establishing a direct connection between the conditions that hold at a local minimum of the penalty function and the conditions that hold at a local solution of (M), as given in Chapter 2.

Returning to the general development, we wish to be able to solve Problem M (equivalently, Problem B), when we require only that $x \in R$ or that $x \in Q$; that is, we wish to solve the problems

$$\text{minimize } f(x) \quad [\text{M}(R)]$$

subject to

$$x \in R$$

and

$$\text{minimize } f(x) \quad [\text{M}(Q)]$$

subject to

$$x \in Q$$

using an auxiliary function technique analogous to that given above for Problem M.

4.4 Generalized Interior and Exterior Point Method

Problems M(R) and M(Q) have the same basic structure, the essential difference being that we assume $R^0 \neq \emptyset$ and $Q \neq \emptyset$ (so that we may have $Q^0 = \emptyset$). The remaining distinction is procedural: we insist on restricting x to R^0 in the course of solving [M(R)], whereas x need not be restricted to Q in the course of solving [M(Q)].

Note that (M) is the same as [M(R)] if $Q = E^n$ and (M) is the same as [M(Q)] if $R = E^n$. Since there were no restrictions on R and Q in the above development (other than $R^0 \neq \emptyset$, $Q \neq \emptyset$), the above convergence theorem for Problem M utilizing $V(x, r, t)$ is valid for $R = E^n$ or $Q = E^n$. We modify the respective definitions of $V(x, r, t)$ in that we associate the parameter r with Problem M(R) and the parameter t with Problem M(Q).

We are thus led to the V functions for [M(R)] and [M(Q)], which, following the development in the previous sections, we call $U(x, r)$ and $T(x, t)$, respectively.

We assume $R^0 \neq \emptyset$ and essentially arrive at the defining properties of $U(x, r)$ by setting $Q = E^n$ and by suppressing t in the definition of $V(x, r, t)$.

DEFINING PROPERTIES OF $U(x, r)$

(a) $U(x, r)$ is continuous for $x \in R^0$, for any $r > 0$.
(b) If $\{x^k\} \subset R^0$ and $x^k \to y \in R - R^0$, where $\|y\| < \infty$, then
$$\lim_{k \to \infty} U(x^k, r) = +\infty$$
for any $r > 0$.
(c) If $\{x^k\} \subset R^0$, $r_k > 0$ for every k, and $(x^k, r_k) \to (y, 0)$, where $\|y\| < \infty$, then $\liminf_{k \to \infty} U(x^k, r_k) \geq f(y)$.
(d) If $y \in R^0$, $r_k > 0$ for every k, and $r_k \to 0$, then $\lim_{k \to \infty} U(y, r_k) = f(y)$.

Assuming $Q \neq \emptyset$, setting $R = E^n$, and suppressing r in the definition of $V(x, r, t)$, we obtain the defining properties of $T(x, t)$.

DEFINING PROPERTIES OF $T(x, t)$

(A) $T(x, t)$ is continuous for $x \in E^n$ for any $t > 0$.
(B) If $\{x^k\} \subset E^n$, $t_k > 0$ for every k, and $(x^k, t_k) \to (y, +\infty)$, where $\|y\| < \infty$, then $\liminf_{k \to \infty} T(x^k, t_k) \begin{cases} = +\infty \text{ if } y \notin Q, \\ \geq f(y) \text{ otherwise.} \end{cases}$
(C) If $y \in Q$, $t_k > 0$ for every k, and $t_k \to +\infty$, then $\lim_{k \to \infty} T(y, t_k) = f(y)$.

As direct consequences of these definitions and the previous results, we obtain methods for solving [M(R)] and [M(Q)].

Corollary 15 [Convergence of U Minima to Local Solutions of Problem M(R)]. The convergence theorem for $V(x, r, t)$ applied to (M) is valid if

$V(x, r, t)$ is replaced by $U(x, r)$, (M) is replaced by [M(R)], $Q = E^n$, and t is suppressed in the statement of that theorem.

Corollary 16 [Convergence of T Minima to Local Solutions of Problem M(Q)]. The convergence theorem for $V(x, r, t)$ applied to (M) is valid if $V(x, r, t)$ is replaced by $T(x, t)$, (M) is replaced by [M(Q)], $R = E^n$, and r is suppressed in the statement of that theorem.

The steps in the proofs of these two theorems are precisely the same as those in the V-function theorem, with obvious modifications, on making the indicated substitutions.

4.5 A HIERARCHY OF PENALTY FUNCTIONS

We obtain a realization of $V(x, r, t)$ in terms of $U(x, r)$ and $T(x, t)$ in the following lemma by defining $V_1(x, r, t) \equiv U(x, r) + T(x, t) - f(x)$. As required by the definition of $V(x, r, t)$, we assume that the problem functions are continuous, $R^0 \neq \emptyset$, and $Q \neq \emptyset$.

Lemma 3. $V_1(x, r, t)$ is a V function [that is, a function $V(x, r, t)$ satisfying (i–iv) above.]

Proof. Property i of $V(x, r, t)$ is an immediate consequence of Property a of $U(x, r)$ and Property A of $T(x, t)$.

To verify Property ii, we require that if $\{x^k\} \subset R^0$, $x^k \to y \in R - R^0$, and $\|y\| < \infty$, then $\lim_{k \to \infty} V(x^k, r, t) = +\infty$ for any $r > 0$, $t > 0$. It follows that $\lim_{k \to \infty} U(x^k, r) = +\infty$ by Property b, for any $r > 0$. By (A) and the fact that $\|y\| < \infty$, it follows that $\lim_{k \to \infty} T(x^k, t) = T(y, t)$, a finite number, for any $t > 0$. By the continuity of $f(x)$ and the fact that $\|y\| < \infty$, $\lim_{k \to \infty} f(x^k) = f(y)$, a finite number. Thus (ii) is satisfied.

To verify (iii), assume $\{x^k\} \subset R^0$, $r_k > 0$, and $t_k > 0$ for every k,

$$(x^k, r_k, t_k) \to (y, 0, +\infty),$$

and $\|y\| < \infty$. By (c), $\liminf_{k \to \infty} U(x^k, r_k) \geq f(y)$, a finite number, using also the fact that $\|y\| < \infty$ and the continuity of $f(x)$. This latter fact also implies that $\lim_{k \to \infty} f(x^k) = f(y)$. Since $\lim_{k \to \infty} T(x^k, t_k) = +\infty$ if $y \notin Q$ by (B), $\lim_{k \to \infty} V_1(x^k, r_k, t_k) = +\infty$ if $y \notin Q$, and so the first part of (iii) holds for $V_1(x, r, t)$. If $y \in Q$, $\liminf_{k \to \infty} T(x^k, t_k) \geq f(y)$ by (B), and so $\liminf_{k \to \infty} V_1(x^k, r_k, t_k) \geq f(y)$ and the second part of (iii) also holds.

Finally, if $y \in R^0 \cap Q$, each term of $V_1(y, r_k, t_k)$ converges to $f(y)$ by (d) and (C) and the continuity of $f(x)$, and so $\lim_{k \to \infty} V_1(y, r_k, t_k) = f(y)$, provided that $r_k > 0$ and $t_k > 0$ for every k, and $(r_k, t_k) \to (0, +\infty)$. Thus (iv) also holds, and so $V_1(x, r, t)$ is a V function. Q.E.D.

4.5 A Hierarchy of Penalty Functions

The U and T functions treated previously are of the form $f(x)$ plus a "penalty" function, which absorbs the effects of the constraints of the given problem and the "control" parameter. Toward developing these functions as particular realizations of the functions $U(x, r)$ and $T(x, t)$ defined above, we define the following. Assume $R^0 \neq \emptyset$ and $Q \neq \emptyset$.

DEFINING PROPERTIES OF $I(x, r)$

(a_1) $I(x, r)$ continuous for $x \in R^0$, for any $r > 0$.
(b_1) If $\{x^k\} \subset R^0$ and $x^k \to y \in R - R^0$, where $\|y\| < \infty$, then
$$\lim_{k \to \infty} I(x^k, r) = +\infty$$
for any $r > 0$.
(c_1) If $\{x^k\} \subset R^0$, $r_k > 0$ for every k and $(x^k, r_k) \to (y, 0)$, where $\|y\| < \infty$, then $\liminf_{k \to \infty} I(x^k, r_k) \geq 0$.
(d_1) If $y \in R^0$, $r_k > 0$ for every k and $r_k \to 0$, then $\lim_{k \to \infty} I(y, r_k) = 0$.

DEFINING PROPERTIES OF $O(x, t)$

(A_1) $O(x, t)$ is continuous for $x \in E^n$, for any $t > 0$.
(B_1) If $\{x^k\} \subset E^n$, $t_k > 0$ for every k and $(x^k, t_k) \to (y, +\infty)$, where $\|y\| < \infty$, then $\liminf_{k \to \infty} O(x^k, t_k) \begin{cases} = +\infty \text{ if } y \notin Q, \\ \geq 0 \text{ otherwise.} \end{cases}$
(C_1) If $y \in Q$, $t_k > 0$ for every k and $t_k \to +\infty$, then $\lim_{k \to \infty} O(y, t_k) = 0$.

With "U function" and "T function" meaning $U(x, r)$ and $T(x, t)$, respectively, as defined in the preceding section, we obtain the following direct consequences.

Lemma 4. $U_1(x, r) \equiv f(x) + I(x, r)$ is a U function.

Lemma 5. $T_1(x, t) \equiv f(x) + O(x, t)$ is a T function.

The proofs of the lemmas follow immediately from the continuity of $f(x)$ and the defining properties given above.

A further immediate consequence of the three lemmas above is the following.

Corollary 17.
$$V_2(x, r, t) \equiv U_1(x, r) + T_1(x, t) - f(x) = f(x) + I(x, r) + O(x, t)$$
is a V function.

This provides a realization of the V function associated with Problem M, in terms of the objective function $f(x)$ of that problem, and the penalty functions associated with Problems M(R) and M(Q) [through $U_1(x, r)$ and $T_1(x, t)$ above].

The functions defined in the previous development (Sections 3.2 and 4.1), applied to Problems M(R), M(Q), and M, are, respectively,

$$U_2(x, r) \equiv f(x) + s(r)I(x),$$

$$T_2(x, t) \equiv f(x) + p(t)O(x),$$

$$V_3(x, r, t) \equiv f(x) + s(r)I(x) + p(t)O(x).$$

For convenience we repeat the defining properties of $s(r)$, $I(x)$, $p(t)$, and $O(x)$, and show that $s(r)I(x)$ is an I function (satisfies Properties a_1 through d_1 above), and $p(t)O(x)$ is an O function (satisfies Properties A_1 through C_1 above). In view of the above results, this in turn means that U_2 is a U function, T_2 is a T function, and V_3 is a V function, as defined above.

DEFINING PROPERTIES OF $I(x)$ AND $s(r)$

(1) $I(x)$ is continuous for $x \in R^0$.
(2) If $\{x^k\} \subset R^0$ and $x^k \to y \in R - R^0$, then $\lim_{k \to \infty} I(x^k) = +\infty$.
(3) $s(r)$ is a (scalar-valued) function of r, continuous for $r > 0$.
(4) If $r_1 > r_2 > 0$, then $s(r_1) > s(r_2) > 0$.
(5) If $r_k > 0$ for every k and $r_k \to 0$, then $\lim_{k \to \infty} s(r_k) = 0$.

Lemma 6. $I_1(x, r) \equiv s(r)I(x)$ is an I function.

Proof. Properties 1 and 3 imply that $I_1(x, r)$ is continuous for $x \in R^0$ and $r > 0$, and so Property a_1 of $I(x, r)$ is satisfied.

For (b_1), assume $\{x^k\} \subset R^0$, $x^k \to y \in R - R^0$, and $r > 0$. Then $s(r) > 0$ by (^{4}) and we can conclude that $I(x^k) \geq M/s(r)$ for any positive number M, provided that k is large enough, by Property 2. This means $s(r)I(x^k) \geq M$, and so $\lim_{k \to \infty} s(r)I(x^k) = +\infty$, since M can be made as large as desired by selecting k large enough. Since this is true for any $r > 0$, Property b_1 holds for $I_1(x, r)$.

To verify (c_1), assume $\{x^k\} \subset R^0$, $r_k > 0$ for every k, $(x^k, r^k) \to (y, 0)$, and $\|y\| < \infty$. Then $s(r_k) > 0$ for every k by (4). If $y \in R - R^0$, $I(x^k) \to +\infty$ by (2), so that for this case $s(r_k)I(x^k) > 0$ for all k large enough. This implies $\liminf_{k \to \infty} s(r_k)I(x^k) \geq 0$, and so ($c_1$) holds for $y \in R - R^0$. The other possibility is that $y \in R^0$. If this is so, $I(x^k) \to I(y)$, a finite number, by (1) and the fact that $\|y\| < \infty$. Since we are assuming $r_k > 0$ and $r_k \to 0$, $s(r^k) \to 0$ by (5). Hence $s(r_k)I(x^k) \to 0$, and consequently Property c_1 is satisfied by $I_1(x, r)$.

Finally, if $y \in R^0$, $I(y)$ is defined and finite by (1), so that if, $r_k > 0$ for every k and $r_k \to 0$, $s(r_k)I(y) \to 0$, by (5); therefore Property d_1 also holds, and hence $I_1(x, r)$ is an I function. Q.E.D.

4.5 A Hierarchy of Penalty Functions

DEFINING PROPERTIES OF $O(x)$ AND $p(t)$

(1^1) $O(x)$ is continuous for $x \in E^n$.

(2^1) $O(x) \begin{cases} = 0 \text{ if } x \in Q, \\ > 0 \text{ if } x \notin Q. \end{cases}$

(3^1) $p(t)$ is a (scalar-valued) function of t, continuous for $t > 0$.
(4^1) If $t_2 > t_1 > 0$, then $p(t_2) > p(t_1) > 0$.
(5^1) If $t_k > 0$ for every k and $t_k \to +\infty$, then $\lim_{k \to \infty} p(t_k) = +\infty$.

Lemma 7. $O_1(x, t) \equiv p(t)O(x)$ is an O function.

Proof. Properties 1^1 and 3^1 imply that $O(x, t)$ is continuous for $x \in E^n$ and $t > 0$, and so Property A_1 of $O(x, t)$ is satisfied by $O_1(x, t)$.

To verify (B_1), assume $\{x^k\} \subset E^n$, $t_k > 0$ for every k, and $(x^k, t_k) \to (y, +\infty)$, where $\|y\| < \infty$. First, since $p(t_k)O(x^k) \geq 0$ for every k by (2^1) and (4^1), $\liminf_{k \to \infty} p(t_k)O(x^k) \geq 0$, and so the second part of (B_1) holds. If $y \notin Q$, $\liminf_{k \to \infty} O(x^k) = O(y) = \delta > 0$ and is finite by (1^1) and (2^1) and the fact that $\|y\| < \infty$. Since $p(t_k) \to +\infty$ by (5^1), this means $p(t_k)O(x^k) \geq (M/\delta)O(x^k)$ for k large enough, where M can be selected as large as desired by choosing k suitably large. Hence $\liminf_{k \to \infty} p(t_k)O(x^k) \geq M$, and $M(k) > 0$, so that $p(t_k)O(x^k) \to +\infty$ and we have also satisfied the first part of (B_1).

Finally, if $y \in Q$ then $O(y) = 0$ by (2^1). Hence with $t_k > 0$ for every k and $t_k \to +\infty$, since $p(t_k)O(y) \equiv 0$ for every k, $p(t_k)O(y) \to 0$, and (C_1) is satisfied.

All the properties of $O(x, t)$ are satisfied by $O_1(x, t)$.   Q.E.D.

5

Extrapolation in Unconstrained Minimization Techniques

Looking at Figures 4 and 5 one might expect that when a set of local minima consists of one point a unique trajectory of unconstrained local minima exists converging to that point. Furthermore, one would hope that an examination of a few points on such a trajectory would give information about the final point, the local minimum to which it is converging.

It is the intention of this chapter to explore trajectories of local unconstrained minima converging to single-point sets of constrained minima under various assumptions on the differentiability of the problem functions. Aside from their intrinsic interest, the results of this chapter have important computational implications, which are discussed further in Section 8.4.

5.1 TRAJECTORY ANALYSIS FOR INTERIOR POINT TECHNIQUES

The assumptions needed to define a unique trajectory of local unconstrained minima are stronger than those needed to prove the existence of points converging to a local minimum. We do not insist literally on uniqueness of the trajectory, but rather on its being "isolated," or locally unique. We shall have occasion to consider the parameter r as a continuous variable for which $U(x, r)$ is to be minimized, rather than confining ourselves to a sequence of discrete values $\{r_k\}$, as has been done up to now.

The meaning of an isolated trajectory is given as follows.

Definition. *A vector function $x(r)$ defined on $(0, r_0]$ is an isolated trajectory of unconstrained local minima of $U(x, r)$ in R^0 if $x(r)$ is continuous and if $x(\bar{r})$ is an isolated unconstrained local minimum of $U(x, \bar{r})$ for any $\bar{r} \in (0, r_0]$*

5.1 Trajectory Analysis for Interior Point Techniques

We also modify our requirements on $I(x)$ and $s(r)$. Rather than simply being a function of x (see Section 3.2), I is now defined as a function of x through the constraint functions $g \equiv (g_1, \ldots, g_m)$. Let $I(g) = \sum_{i=1}^{m} I_i(g_i)$ be a twice-differentiable function of the g_i when each $g_i > 0$ with the following properties. If $\{g_i^k\}$ is an infinite sequence of points where $g_i^k > 0$ for each k, and $\lim_{k \to \infty} g_i^k = 0$, then $\lim_{k \to \infty} I_i(g_i^k) = +\infty$. Furthermore, if $g_i^0 > 0$, then $\partial I_i(g_i^0)/\partial g_i < 0$, $\partial^2 I_i(g_i^0)/\partial g_i^2 > 0$, and $\partial^2 I_i/\partial g_i^2$ is a monotonically *decreasing* function of g_i.

For simplicity in presentation the generality in the function $s(r)$ will be dropped and we shall let $s(r) = r$.

We note that these new requirements on $I[g(x)]$ and $s(r)$ satisfy the properties given in Section 3.2 for U to be an interior point unconstrained minimization function, and therefore all theorems proved thus far remain valid.

Theorem 12 [Existence of an Isolated Trajectory]. If (a) the functions f and $\{g_i\}$ are twice differentiable, (b) at x^* there exists a u^* such that the *sufficient* conditions (2.26–2.31) for x^* to be a constrained local minimum of Problem B are satisfied, and $B^* = D^*$, (c) the vectors ∇g_i (all $i \in B^*$) are linearly independent, and (d) $I(g)$ satisfies the requirements just stated, then

(i) x^* is an *isolated* local constrained minimum of Problem B [let $v^* = f(x^*)$],

(ii) the variables u_i^*, $i = 1, \ldots, m$ are unique and are generated explicitly by any interior point unconstrained minimization technique (see Section 3.2), and

(iii) there exists an isolated, once-differentiable trajectory $x(r)$ of local unconstrained minima of $U(x, r) \equiv f(x) + rI[g(x)]$ (defined in Section 3.2) in R^0 converging to x^*.

Proof. Part i is a statement of what was proved in Theorem 4. The independence of the gradients of all binding constraints yields the result that $x^* \in \overline{R^0}$ (that is, there are points in R^0 arbitrarily close to x^*). Hence all the hypotheses of Corollary 10, Section 3.3, are satisfied, and there exists (for r small enough) at least one function $x(r)$ of unconstrained local minima converging to x^*. That a function exists that defines an isolated trajectory (that is, the function is continuous and every point on the trajectory is an isolated, unconstrained minimum) will be proved in several stages.

(1) The u_i^* for which

$$\nabla f(x^*) - \sum_{i \in B^*} u_i^* \nabla g_i(x^*) = 0$$

as stated in Theorem 4 are *unique*. This follows directly from the assumption that the ∇g_i^* are independent for $i \in B^*$. In fact, $u^* = (G^T G)^{-1} G^T \nabla f^*$, where $G = \{\nabla g_i\}, i \in B^*$.

(2) At every $x(r_k)$

$$\nabla_x U[x(r_k), r_k] = \nabla_x f[x(r_k)] + \nabla_x I\{g[x(r_k)]\} r_k = 0. \tag{5.1}$$

This follows because $x(r_k) \in R^0$ is an unconstrained local minimum of $U(x, r_k)$, the problem functions are differentiable in x, and I is differentiable in g for $g > 0$, and also from (2.12). Using the chain rule,

$$\nabla_x f[x(r_k)] + \nabla_x g[x(r_k)] \nabla_g I\{g[x(r_k)]\} r_k = 0. \tag{5.2}$$

(3) Define $u_i^k = -[\partial I_i\{g_i[x(r_k)]\}/\partial g_i] r_k$ for $i = 1, \ldots, m$. Then the limit of u_i^k as $r_k \to 0$ is *unique* and equal to u_i^*, $i = 1, \ldots, m$.

To prove this, note that (5.2) can be written

$$\nabla f^k - \sum_{i=1}^m u_i^k \nabla g_i^k = 0 \quad \text{[evaluated at } x(r_k)\text{]}. \tag{5.3}$$

Clearly for all i, where $g_i^* > 0$, $\lim_{k \to \infty} u_i^k$ exists and is equal to zero.

Let $d^k = \sum_{i=1}^m u_i^k$. Now if $m \geq 1$, then $d^k > 0$. (If $m = 0$, we are dealing with an unconstrained problem and Part 3 holds trivially.) Let

$$v_i^k = \frac{u_i^k}{d^k}, \quad i = 1, \ldots, m.$$

Let $\bar{d} = \lim \inf_{k \to \infty} d^k$. If $\bar{d} = +\infty$, then dividing (5.3) by d^k and taking the limit as $k \to \infty$ yields

$$\sum_{i \in B^*} \bar{v}_i \nabla g_i^* = 0$$

for a set of non-negative $\bar{v}_i$ where $\sum_{i \in B^*} \bar{v}_i = 1$. But this contradicts the independence of the ∇g_i^*'s (all $i \in B^*$). If $\bar{d} < +\infty$, then let u^0 represent any point of accumulation of the sequence $\{u^k\}$. Then from (5.3)

$$\nabla f^* = \sum_{i \in B^*} u_i^0 \nabla g_i^*.$$

By the independence of the ∇g_i^*'s, $u_i^0 = u_i^*$ (all $i \in B^*$). This completes the proof of Part ii.

(4) For all i where $u_i^* > 0$,

$$\lim_{k \to \infty} \inf r_k \frac{\partial^2 I_i g_i[x(r_k)]}{\partial g_i^2} = +\infty.$$

For shorthand let $g_i[x(r_k)]$ be written as g_i^k and use analogous notation for the other functions. Let $l = l(k) < k$ be an index available for every k so that $r_l > 2 r_k$, $g_i^l > g_i^k$, and $\lim_{k \to \infty} r_l = 0$. Since $\{r_k\}$ is a decreasing

5.1 Trajectory Analysis for Interior Point Techniques

null sequence, such an index is always available when k is large enough. Now

$$r_k\left[\frac{\partial I_i^l}{\partial g_i} - \frac{\partial I_i^k}{\partial g_i}\right] \leq r_k[g_i^l - g_i^k]\frac{\partial^2 I_i^k}{\partial g_i^2}$$

using the monotonic property on $\partial^2 I_i/\partial g_i^2$. Also

$$r_l/2(\partial I_i^l/\partial g_i) < r_k\, \partial I_i^l/\partial g_i \quad (r_l > 2r_k,\ \partial I_i^l/\partial g_i < 0).$$

Hence

$$0 < \frac{u_i^*}{2} = \liminf_{k\to\infty} \tfrac{1}{2}r_l\frac{\partial I_i^l}{\partial g_i} - r_k\frac{\partial I_i^k}{\partial g_i} \quad [\text{recall } l = l(k)]$$

$$\leq \liminf_{k\to\infty} [g_i^l - g_i^k]r_k\frac{\partial^2 I_i^k}{\partial g_i^2}.$$

Part 4 follows from the last inequality.

(5) For r_k small enough the matrix $\nabla^2 U[x(r_k), r_k]$ is *positive definite*. Using the chain rule,

$$\nabla^2 U^k = \nabla^2 f^k - \sum_{i=1}^m u_i^k \nabla^2 g_i^k$$
$$+ \sum_{i=1}^m \nabla g_i^k \frac{\partial^2 I_i(g_i^k)}{\partial g_i^2} r_k \nabla^T g_i^k. \tag{5.4}$$

Consider any sequence $0 < r_k \to 0$ and $z_k \in E^n$ such that $\|z_k\| = 1$ for every k. It suffices to show that $z_k^T \nabla^2 U^k z_k > 0$ for k large. Select a convergent subsequence of $\{z_k\}$, relabel it $\{z_k\}$ for convenience, and call the limit (unit vector) $\bar{z}$. If $\bar{z}^T \nabla g_i^* \neq 0$ for some i such that $u_i^* > 0$, then $z_k^T \nabla^2 U^k z_k \to +\infty$ as $k \to \infty$ because the quantity derived from the third term of (5.4), $|z_k^T \nabla g_i^k|^2 r_k[\partial^2 I_i(g_i^k)/\partial g_i^2] \to +\infty$ as $k \to \infty$ and dominates other terms.

For those vectors $\bar{z}$ where $\bar{z}^T \nabla g_i^* = 0$ for all i such that $u_i^* > 0$, since the sufficiency conditions of Theorem 4 are assumed to hold, then for r_k small enough (k large enough)

$$z_k^T \nabla^2 U^k z_k \geq z_k^T [\nabla^2 f^k - \sum_{i=1}^m u_i^k \nabla^2 g_i^k] z_k > 0.$$

(If all $u_i^* = 0$, then the conditions imply that x^* is an unconstrained local minimum of $f(x)$, since we would have $\nabla^2 f(x^*)$ positive definite. Again we would obtain $z_k^T \nabla^2 U^k z_k > 0$ for large k.) This proves Part 5.

(6) For r_k small enough, any $x(r_k)$ is an *isolated* local unconstrained minimum of $U[x, r_k]$. This follows directly from Part 5, (5.3), and Corollary 6, Section 2.3.

(7) For r_k small enough, about any $x(r_k)$ there is a unique, once-differentiable function $x(r)$ such that each $x(r^0)$ is an isolated local unconstrained minimum of $U(x, r^0)$. Furthermore $x(r)$ is defined for $r_k \geq r > 0$.

The existence of $x(r)$ in a neighborhood about r_k can be proved from the *implicit function theorem* as follows. At $x(r_k)$, from (5.3),

$$\nabla f[x(r_k)] - \sum_{i=1}^{m} u_i^k \nabla g_i[x(r_k)] = 0. \tag{5.5}$$

This is a system of n equations in $n + 1$ unknowns. The Jacobian matrix of (5.5) with respect to x is the matrix in (5.4). In (v) it was proved that for r_k small enough $\nabla^2 U^k$ was positive definite and therefore had an inverse. The implicit function theorem [102, page 181] can be used since the Jacobian of (5.5) does not vanish. Then in a neighborhood about r_k there is a unique, once-differentiable function $x(r)$ passing through the chosen $x(r_k)$ such that

$$\nabla f[x(r)] - \sum_{i=1}^{m} u_i(r) \nabla g_i[x(r)] = 0.$$

For r close to r_k, $x(r)$ is close to $x(r_k)$, $\nabla^2 U[x(r), r]$ is positive definite, and $x(r)$ defines a unique, continuously differentiable trajectory of isolated local minima. Assume that $x(r)$ does not define local minima for all $r_k \geq r > 0$. Let $r^0 = \inf r$, for which $x(r)$ describes a trajectory of local unconstrained minima. Let $x(r^0)$ be an accumulation point of $x(r)$ as $r \to r^0$. One exists since $x(r)$ is contained in the compact S set given by Theorem 7. Since the functions involved are continuous, clearly (5.3) holds at $x(r^0)$. That $\nabla^2 U[x(r^0), r^0]$ is positive definite can be seen by assuming that r_k was small enough so that for all unit vectors z such that $z^T \nabla g_i^* = 0$ (for all i where $u_i^* > 0$), $z^T[\nabla^2 f^k - \sum u_i^k \nabla^2 g_i^k]z > \epsilon_1 > 0$, and such that $z^T \nabla^2 U^k z$ is dominated by the third term of (5.4) for all other z. Clearly, then, $\nabla^2 U[x(r^0), r^0]$ is positive definite, proving that $x(r^0)$ is also an isolated local unconstrained minimum. Again a neighborhood exists about r^0 and the function $x(r)$ can be extended and defined for r in this neighborhood so that r^0 is not the infimum of all $0 < r \leq r_k$, for which $x(r)$ defines an isolated, once-differentiable trajectory of unconstrained minima. This contradicts our assumption and we have shown an isolated trajectory exists for all $r_k \geq r > 0$.   Q.E.D.

5.2 ANALYSIS OF ISOLATED TRAJECTORY

Having established that under certain assumptions there is an isolated trajectory of points converging to an isolated local unconstrained minimum, we now show that under the same assumptions this trajectory has an order of differentiability with respect to the variable r (when $r > 0$) one less than that of the original problem functions and that it is analytic when the functions are analytic. Furthermore we prove that with respect to r, limiting derivatives

5.2 Analysis of Isolated Trajectory

exist at $r = 0$ if all the dual variables associated with the binding constraints are strictly positive. From now on it will be assumed that we are talking about unconstrained minima on one isolated trajectory.

It is possible to be explicit about the derivatives of $x(r)$ with respect to r for $r > 0$. Since (5.1) is an identity in r we can differentiate, obtaining

$$\nabla^2 U[x(r), r] \, Dx(r) + \nabla I[x(r)] = 0. \tag{5.6}$$

Under the assumptions of Theorem 11, the matrix that multiplies $Dx(r)$ in (5.6) has an inverse, so that

$$Dx(r) = -\{\nabla^2 U[x(r), r]\}^{-1} \nabla I[x(r)].$$

Then the derivative of $x(r)$ with respect to r exists for $r > 0$. If we differentiate (5.6) with respect to r,

$$\nabla^2 U[x(r), r] \, D^2x(r) + \frac{d\nabla^2 U[x(r), r]}{dr} Dx(r) + \nabla^2 I[x(r)] \, Dx(r) = 0.$$

The existence of the same inverse is required for $D^2x(r)$ to exist as required for $Dx(r)$. In addition, $D^2x(r)$ requires the existence of the third partial derivatives of f and g with respect to x, and the third partial derivatives of I with respect to $g_1, \ldots, g_m$.

By continuing in this manner it is possible to obtain explicitly all derivatives $D^k x(r)$ in terms of the derivatives $D^i x(r)$ ($i = 1, \ldots, k - 1$), and partial derivatives of the problem functions of degree up to $k + 1$. By induction, the following theorem is proved.

Theorem 13 [Existence of $D^k x(r)$ for $r > 0$]. If, in addition to the hypotheses of Theorem 12, the functions f and $\{g_i\}$ have continuous partial derivatives with respect to x of degree $k + 1$ ($k + 1 \geq 2$) and I has continuous partial derivatives with respect to $g_1, \ldots, g_m$ of degree $k + 1$, then

$$D^i x(r), \quad i = 1, \ldots, k$$

exists and can be explicitly obtained by repeated differentiation of (5.6).

One possible use of this theorem would be to attempt to approximate the solution $x^* = x(0)$ by using a finite Taylor's series approximation

$$x(0) \doteq \sum_{i=1}^{p} \frac{D^i[x(r_1)](r_1)^i(-1)^i}{i!}. \tag{5.7}$$

Just minimizing for one value of $r = r_1$ and using the iterative technique for generating successive derivatives $D^i[x(r_1)]$ might enable one to approximate the solution by using the power series expansion (5.7).

A result in this direction is given in the next corollary.

Corollary 18 [Analyticity of $x(r)$ for $r > 0$]. If in addition to the hypothesis of Theorem 12 the functions f and $\{g_i\}$ are analytic and I is an analytic function of g, then about any $r_1 > 0$ there exists a unique function $x(r)$ that is analytic in a neighborhood of r_1, where $x(r)$ is an isolated trajectory of unconstrained minima of $U(x, r)$.

Proof. The corollary follows from (a) the assumption that the Jacobian of $\nabla^2 U[x(r_1), r_1]$ does not vanish, (b) the analyticity of the functions involved, and (c) the implicit function theorem for real or complex variables. (See Bochner and Martin [12, Theorem 8, page 39].)

This corollary is still of limited use, since there is no guarantee that for r_1 small enough the domain of analyticity of $x(r)$ about r_1 includes $r = 0$. The conditions that ensure this are stated in Corollary 19.

We show how the solution to the example of Figure 3, Section 3.2, can be estimated by utilizing only one unconstrained minimization using (5.7). Clearly the problem functions are infinitely differentiable. The derivatives at $r = 0.055$ are given along with the estimates based on them.

Example [Solution by Power Series]. Recall for the example of Figure 3

$$x_1(r) = \frac{-1 + \sqrt{1 + 8r}}{4} = 0.0500 \quad (r = 0.055),$$

$$x_2(r) = \frac{(-1 + \sqrt{1 + 8r})^2}{16} + r = 0.0575 \quad (r = 0.055).$$

Derivative	Expression	Numerical Value	Solution Estimate
$Dx_1(r)$	$(1 + 8r)^{-1/2}$	0.833333	0.0041766
$Dx_2(r)$	$2x_1 Dx_1 + 1$	1.083333	−0.00208
$D^2 x_1(r)$	$-4(1 + 8r)^{-3/2}$	−2.314815	0.000665
$D^2 x_2(r)$	$2(Dx_1)^2 + 2x_1 D^2 x_1$	1.1573	−0.00033
$D^3 x_1(r)$	$48(1 + 8r)^{-5/2}$	19.290123	0.000114
$D^3 x_2(r)$	$6(Dx_1)D^2 x_1 + 2x_1 D^3 x_1$	−9.643745	−0.000063

Clearly the estimates of the optimum are converging as more derivatives are used. The convergence is not rapid. If we had started with a smaller value of r, the estimates would have converged more rapidly. (Quite obviously, the example of Figure 4 can be solved using just one estimate.)

Analysis of Logarithmic Trajectory at $r = 0$

We have shown that under appropriate assumptions isolated trajectories with differing orders of differentiability (with respect to r) exist converging to isolated local minima. The next question is whether or not the limits of these derivatives exist at $r = 0$. One additional hypothesis is needed to prove that these limits are finite. When they are finite, it is possible to develop a scheme based on using the values of $x(r)$ along its trajectory to estimate x^* very accurately. (See Section 8.4 for computational implications.)

Although it is possible to analyze these limits for *any* interior point unconstrained minimization function, the precise choice of r^q for which $d[x(0)]/dr^q$ (and all following derivatives) is finite depends on the particular choice of $I(x)$. We now let $I(x) = -\sum_{i=1}^{m} \ln g_i(x)$. Our unconstrained minimization function is now the following.

Definition.

$$L(x, r) = f(x) - r \sum_{i=1}^{m} \ln g_i(x). \tag{5.8}$$

Recalling from the results of Theorem 12 the two identities in r that hold at $x(r)$, for logarithmic penalty function (5.8), we have

$$\nabla f[x(r)] - \sum_{i=1}^{m} u_i(r) \nabla g_i[x(r)] = 0 \tag{5.9}$$

and

$$u_i(r) g_i[x(r)] = r, \quad i = 1, \ldots, m. \tag{5.10}$$

[For different forms of $I(x)$, (5.10) must be modified. For the P function (3.10) the identity $u_i^{1/2}(r) g(r) = r$ would be substituted, and the derivatives with respect to r taken in a manner similar to what follows.] We regard both x and u as separate functions of r. Differentiating with respect to r yields

$$\begin{bmatrix} \nabla^2 f - \sum_{i=1}^{m} u_i \nabla^2 g_i, & -\nabla g_1, \ldots, -\nabla g_m \\ u_1 \nabla^T g_1 & g_1 \\ \cdot & \cdot \\ \cdot & \cdot \\ \cdot & \cdot \\ u_m \nabla^T g_m & g_m \end{bmatrix} \begin{bmatrix} Dx(r) \\ \\ \\ Du(r) \\ \\ \end{bmatrix} = \begin{bmatrix} 0 \\ 1 \\ \cdot \\ \cdot \\ \cdot \\ 1 \end{bmatrix}, \tag{5.11}$$

where the matrix in (5.11) is evaluated at $[x(r), u(r)]$. Note that we showed $Dx(r) \in C$ for $r > 0$. To prove the $\lim_{r \to 0+} [Dx(r), Du(r)]$ exists it is sufficient to prove that the $m + n$ equations (5.9) and (5.10) in (x, u, r) at $r = 0$ are uniquely satisfied by (x^*, u^*); that is, the Jacobian matrix of (5.9) and (5.10) with respect to the $m + n$ variables, components of (x, u), has an inverse at $r = 0$. That matrix evaluated at x^*, u^* is the same as the limiting matrix in (5.11) as $r \to 0$. Thus we require the following theorem.

Theorem 14 [Existence of $Dx(0), Du(0)$]. If (a) the functions f and $\{g_i\}$ are twice differentiable, (b) the gradients $\{\nabla g_i^*\}$ (all $i \in B^*$) are linearly independent, (c) strict complementarity holds for

$$u_i^* g_i(x^*) = 0, \qquad i = 1, \ldots, m$$

[that is, $u_i^* > 0$ if $g_i(x^*) = 0$], and (d) the sufficient conditions (2.26–2.31) that x^* be a local constrained minimum of Problem B are satisfied by (x^*, u^*), then there is a neighborhood about $r = 0$ for which a unique, continuously differentiable function $[x(r), u(r)]$ exists satisfying (5.9) and (5.10) for which (when $r > 0$) $x(r)$ describes a unique isolated trajectory of local minima of $L(x, r)$, $x(r) \to x^*$, and $u(r) \to u^*$.

Proof. We need only show that the matrix in (5.11) is nonsingular at (x^*, u^*). Then the implicit function theorem [102, page 181] can be applied to give the differentiable function $[x(r), u(r)]$. Since it is unique and satisfies (5.9) and (5.10), $x(r)$ must be the isolated trajectory (when $r > 0$) whose existence was proved in Theorem 12. That is, under the assumption of strict complementarity, there is only one isolated trajectory converging to x^*. To prove the existence of the inverse we need only show that there is no nonzero solution of the equation

$$\begin{bmatrix} \nabla^2 f^* - \sum_{i=1}^{m} u_i^* \nabla^2 g_i^*, & -\nabla g_1^*, \ldots, -\nabla g_m^* \\ u_1^* \nabla^T g_1^* & g_1^* \\ \vdots & & \ddots \\ u_m^* \nabla^T g_m^* & & & g_m^* \end{bmatrix} \begin{bmatrix} z_1 \\ z_2 \end{bmatrix} = \begin{bmatrix} 0 \\ 0 \end{bmatrix}. \qquad (5.12)$$

Clearly $z_2{}^i = 0$ for all i such that $g_i^* > 0$. Also

$$u_i^* \nabla^T g_i^* z_1 = 0 \quad \text{for all} \quad i \quad \text{such that} \quad g_i^* = 0 \quad (\text{all } i \in B^*). \qquad (5.13)$$

But since strict complementarity is assumed to hold, (5.13) implies $\nabla^T g_i^* z_1 = 0$ (all $i \in B^*$). Clearly

$$\nabla^2 f^* - \sum_{i=1}^{m} u_i^* \nabla^2 g_i^* = \nabla^2 f^* - \sum_{i \in B^*} u_i^* \nabla^2 g_i^*.$$

5.2 Analysis of Isolated Trajectory

Premultiplying (5.12) by $[z_1^T, z_2^T]$ yields

$$z_1^T\left[\nabla^2 f^* - \sum_{i\in B^*}\nabla^2 g_i^*\right]z_1 + \sum_{i=1}^{m}(z_2^i)^2 g_i^* = 0.$$

But when $g_i^* = 0$, $(z_2^i)^2 g_i^* = 0$, and, from above, $z_2^i = 0$ for $g_i^* > 0$. Thus

$$z_1^T\left[\nabla^2 f^* - \sum_{i\in B^*} u_i^* \nabla^2 g_i^*\right]z_1 = 0. \tag{5.14}$$

But since z_1^T is orthogonal to all ∇g_i^*, where $i \in B^*$, from (2.31)

$$z_1^T \nabla^2 \mathcal{L}(x^*, u^*) z_1 > 0$$

if $z_1 \neq 0$. Thus $z_1 = 0$.
Then (5.12) yields

$$\sum_{i\in B^*} z_2^i \nabla g_i^* = 0. \tag{5.15}$$

Since the ∇g_i^*'s for $i \in B^*$ were assumed independent, $z_2^i = 0$ for all $i \in B^*$. But above it was shown that $z_2^i = 0$ for all $i \notin B^*$. Thus $z_1 = 0$ and $z_2 = 0$ for all solutions of (5.12). Q.E.D.

One final theorem whose proof depends on the existence of the inverse matrix in (5.12) follows the pattern of the proof of Theorem 14.

Theorem 15. [Existence of $D^k x(r)$, $D^k u(r)$]. If in addition to the hypotheses of Theorem 14 the differentiability of the problem functions f and $\{g_i\}$ is assumed to be of degree $k + 1 \geq 2$, then the functions $[x(r), u(r)]$ have derivatives of order k in a neighborhood about $r = 0$.

The need for strict complementarity to imply the existence of $dx(0)/dr$ (and all following derivatives) is illustrated by the following example.

Example.

$$\text{minimize } x_2$$

subject to

$$-x_1^2 + x_2 \geq 0,$$
$$x_1 \geq 0.$$

The solution is at $x^* = (0, 0)^T$, and $u_1^* = 1$, $u_2^* = 0$. Hence strict complementarity does not hold. The logarithmic penalty function applied to this is

$$L(x, r) = x_2 - r \ln[-x_1^2 + x_2] - r \ln x_1.$$

It is easy to verify by differentiating that the trajectory leading to the solution is $x_1(r) = (r/2)^{1/2}$ and $x_2(r) = (\frac{3}{2})r$. Thus, although $d^i x(r_1)/dr^i$ exists for all $r_1 > 0$, the limiting value as $r_1 \to 0$ is $+\infty$.

82 Extrapolation in Unconstrained Minimization Techniques

The analyticity of the logarithmic function holds at $r = 0$ under these same conditions when the problem functions are analytic.

Corollary 19 [Analyticity of $x(r)$ at $r = 0$]. If in addition to the hypotheses of Theorem 14 the functions f and $\{g_i\}$ are real analytic, there is a unique analytic function $x(r)$ in a neighborhood about $r = 0$, which, for $r > 0$, defines an isolated trajectory of unconstrained local minima converging to $x^* = x(0)$.

Proof. The proof follows from the existence of the inverse to the matrix in (5.12), the analyticity of the functions involved, and the appropriate form of the implicit function theorem referenced in Corollary 18.

5.3 ANALYSIS OF ISOLATED TRAJECTORY OF EXTERIOR POINT ALGORITHM

Most of the development of Sections 5.1 and 5.2 can be paralleled for exterior point algorithms, and it would be redundant to do so here. Instead we show that for a *particular* function, namely the quadratic loss function defined in Section 4.1, under the same conditions of Theorem 14, a *unique* isolated trajectory converges to an isolated local minimum. The once differentiability and possible analyticity of the trajectory follows from the same conditions.

Theorem 16 [Existence of Isolated Trajectory for Quadratic Loss Function]. Under the assumptions of Theorem 14 there is a unique differentiable function $x(r)$ (for r small) that describes a trajectory of isolated local minima of the quadratic loss unconstrained minimization function

$$f(x) + \sum_{i=1}^{m} \frac{\{[g_i(x) - |g_i(x)|]/2\}^2}{r}$$

converging to an isolated constrained minimum x^*. [Here $p(t) = t = 1/r$, and $t \to \infty$ as $r \to 0$.]

Proof. Assume that the constraints are reordered so that $i \in B^*$, $i = 1, \ldots, \beta$, and $i \notin B^*$, for $i = \beta + 1, \ldots, m$. That is, the binding ones are numbered sequentially from 1 to β. Also u will be treated as a $\beta \times 1$ vector. Then the following two conditions (at $r = 0$) are the analogs of (5.9) and (5.10) under the assumptions of this theorem:

$$\nabla f^* - \sum_{i=1}^{\beta} u_i^* \nabla g_i^* = 0, \tag{5.16}$$

$$\frac{g_i^*}{u_i^*} + \frac{r}{2} = 0, \quad i = 1, \ldots, \beta. \tag{5.17}$$

5.3 Analysis of Isolated Trajectory of Exterior Point Algorithm

Equations 5.16 and 5.17 are $n + \beta$ equations in $n + \beta + 1$ unknowns. The Jacobian matrix with respect to (x, u) is

$$J = \begin{bmatrix} \nabla^2 f^* - \sum_{i=1}^{\beta} u_i^* \nabla^2 g_i^*, & -\nabla g_1^*, \ldots, -\nabla g_\beta^* \\ \dfrac{\nabla^T g_1^*}{u_1^*} & \dfrac{-g_1^*}{(u_i^*)^2} \\ \vdots & \vdots \\ \dfrac{\nabla^T g_\beta^*}{u_\beta^*} & \dfrac{-g_\beta^*}{(u_\beta^*)^2} \end{bmatrix} \quad (5.18)$$

Clearly, by the same reasoning as that used for the proof of Theorem 14, this has an inverse for x^*, u^* when $r = 0$ in (5.16) and (5.17). [Note that in a small neighborhood about x^*, $g_i(x) > 0$ for $i = \beta + 1, \ldots, m$, and these remarks apply without further regard for these constraints.] The implicit function theorem can be applied, and there exist unique continuously differentiable functions $[x(r), u(r)]$ for which (5.16) and (5.17) are satisfied for r close to zero. [We are concerned with r positive since, from (5.17) and the fact that $u_i^* > 0$, for r small, the g_i's are strictly negative as prescribed by exterior functions.] We need to show that the unique function $[x(r)]$ describes a trajectory of unconstrained local minima of the quadratic loss function. To see this, note that the u_i can be solved from (5.17) and

$$u_i(r) = \frac{2g_i[x(r)]}{r}, \quad i = 1, \ldots, \beta. \quad (5.19)$$

Substituting this in (5.16) yields

$$\nabla f[x(r)] + \sum_{i=1}^{\beta} \frac{2g_i[x(r)]}{r} \nabla g_i[x(r)] = 0. \quad (5.20)$$

Thus $x(r)$ satisfies the first-order necessary condition that

$$f(x) + \sum_{i=1}^{\beta} \frac{[(g_i(x) - |g_i(x)|)/2]^2}{r} \quad (5.21)$$

be minimized at $x(r)$. The second partial matrix of (5.21) is

$$\nabla^2 f[x(r)] + \sum_{i=1}^{\beta} \frac{2g_i[x(r)]}{r} \nabla^2 g_i[x(r)] + \sum_{i=1}^{\beta} \nabla g_i[x(r)] \frac{2}{r} \nabla^T g_i[x(r)]. \quad (5.22)$$

For r small, because of our assumptions, this is a positive definite matrix.

Thus the first and second-order sufficient conditions are satisfied by $x(r)$ that it be a local unconstrained minimum of (5.21).   Q.E.D.

Proceeding as in the previous section, all derivatives of $x(r)$ with respect to r can be obtained by repeatedly differentiating (5.20). The theorems regarding the differentiability and analyticity of $x(r)$ for r small follow from the obvious additional assumptions on the problem functions f and $\{g_i\}$ and from using the general form of the implicit function theorem. Thus for r small, extrapolation and acceleration methods can be applied to exterior point algorithms.

5.4 TRAJECTORY ANALYSIS IN MIXED INTERIOR POINT–EXTERIOR POINT ALGORITHMS

If some of the constraints of Problem B were treated by the logarithmic interior point method and some handled by the quadratic loss exterior point method, the conditions under which an isolated trajectory would converge to an isolated point and under which this trajectory would have a derivative with respect to r at $r > 0$ and $r = 0$ would be the same as those in the last two sections. The proofs would be a repetition of the proofs contained in Theorems 12, 13, 14, and 15. Another interesting question is: Under what assumptions does a mixed algorithm applied to the inequality-equality-constrained problem

$$\text{minimize } f(x) \qquad \text{(A)}$$

subject to

$$g_i(x) \geq 0, \quad i = 1, \ldots, m,$$
$$h_j(x) = 0, \quad j = 1, \ldots, p$$

yield the desired trajectories? For definiteness we define the logarithmic-quadratic loss penalty function $W(x, r)$ as

$$W(x, r) = f(x) - r \sum_{i=1}^{m} \ln g_i(x) + \sum_{j=1}^{p} \frac{h_j^2(x)}{r}. \qquad (5.23)$$

[Recall from Section 4.3 that the quadratic loss function applied to two inequality constraints $h_j(x) \geq 0$ and $-h_j(x) \geq 0$ yields one term of the form $h_j^2(x)/r$.]

Theorem 17 [Isolated Trajectory for W Function]. If (a) the functions f, $\{g_i\}$, and $\{h_j\}$ are twice differentiable, (b) the gradients $\{\nabla g_i^*\}$ (all $i \in B^*$) and $\{\nabla h_i^*\}$ ($j = 1, \ldots, p$) are linearly independent, (c) strict complementarity holds for $u_i^* g_i(x^*) = 0$, $i = 1, \ldots, m$, and (d) the sufficient conditions (2.26–2.31) that x^* be an isolated local constrained minimum of Problem A are satisfied by (x^*, u^*, w^*), then there is a positive neighborhood about $r = 0$ for which a unique isolated differentiable function $x(r)$ exists that

5.4 Trajectory Analysis in Mixed Interior–Exterior Algorithms

describes a unique isolated trajectory of local minima of $W(x, r)$, where $x(r) \to x^*$ as $r \to 0$.

Proof. We write the first-order conditions that are assumed to hold at (x^*, u^*, w^*), where $r = 0$:

$$\nabla f - \sum_{i=1}^{m} u_i \nabla g_i + \sum_{j=1}^{p} w_j \nabla h_j = 0, \quad (5.24)$$

$$u_i g_i(x) - r = 0, \quad i = 1, \ldots, m, \quad (5.25)$$

$$2h_j(x) - r w_j = 0, \quad j = 1, \ldots, p. \quad (5.26)$$

There are $n + m + p$ equations in $n + m + p + 1$ variables. The Jacobian matrix of (5.24–5.26) with respect to (x, u, w) is

$$\begin{bmatrix}
\nabla^2 \mathcal{L}(x, u, w), & -\nabla g_1, \ldots, -\nabla g_m, & \nabla h_1, \ldots, \nabla h_p \\
u_1 \nabla g_1^T & g_1 & \\
\vdots & \vdots & \\
u_m \nabla g_m^T & g_m & \\
2\nabla h_1^T & & \\
\vdots & & \text{diag}(-r) \\
2\nabla h_p^T & &
\end{bmatrix} \quad (5.27)$$

Because of our assumptions, the matrix in (5.27) has an inverse at $(x, u, w) = (x^*, u^*, w^*)$, $r = 0$, and hence by the implicit function theorem there is a unique differentiable function $[x(r), u(r), w(r)]$ in a neighborhood about $r = 0$ (we consider only values of $r > 0$) satisfying (5.24–5.26). (The existence of the inverse is proved exactly as in Theorem 14.)

That $x(r)$ describes an isolated trajectory of unconstrained minima to $W(x, r)$ can be seen as follows.

Solving for $u_i(r) = r/g_i[x(r)]$ from (5.25) and $w_j(r) = 2h_j[x(r)]/r$ from (5.26) and substituting in (5.24) yields the first-order condition that $x(r)$ minimize W. Using the method of proof for Theorem 12 it is easily shown that $\nabla^2 W[x(r), r]$ is a positive definite matrix where r is small and hence that $x(r)$ (uniquely) locally minimizes the W function. Q.E.D.

The main use of these theorems, and a very practical one from the standpoint of computer implementation, is that several values of $x(r)$ along the trajectory are used to estimate the solution and thereby accelerate convergence. The formulas required for this are discussed in Section 8.4.

6

Convex Programming

In all the development thus far the main restriction on the problem functions was the order of their differentiability. The theorems that were proved were about "local minima" since it was recognized that in general, given information at a point, one can only make statements about what is true in a neighborhood about that point. For an important class of problems, called *convex programming problems*, local information is also global information.

Convex programming problems have the very desirable property that local minima are global minima. Necessary and sufficient optimality conditions are easily established. An interesting and useful theory of duality has been developed for convex programming. In this chapter it is shown that all the desirable properties transfer to unconstrained minimization algorithms. Local *unconstrained* minima become global *unconstrained* minima, global convergence is guaranteed, and most of the theorems proved "for r_k small" can be shown true "for all $r_k > 0$" in convex programming problems. Important results involving duality are obtained.

It is important to clarify and emphasize the specific role of the convexity assumptions that will be made. We stress that convexity assumptions have not been invoked for any of the results obtained up to this point. Several important additional results of convexity are the following.

1. It implies that a local solution x^* is a global solution of (A).
2. It easily leads to convexity of the function used to transform the problem into a sequence of unconstrained minimizations, subject to some additional restrictions on the penalty term. For example, the interior point function $U[x, r_k]$ is convex, with $f(x)$ and the $\{-g_i(x)\}$ convex, given the appropriate choice of $I(x)$, such as that given in Lemma 11 of this chapter. This leads to the fact that any minimizing point $x(r_k)$ of $U(x, r_k)$ in R^0 is a global minimum.
3. For the interior point methods it leads to the guarantee that $f[x(r_k)]$ decreases monotonically in R as $k \to \infty$. For the exterior point technique it

makes it generally possible to construct a sequence $\{x^k\}$ of feasible points such that $f(x^k)$ will converge to the optimum from above.

4. It leads to valuable tie-ins with duality theory, these being particularly exploitable by the methods given here. The calculation of upper and lower bounds on the optimum value of the programming problem is one of the most important and practical instances of this.

6.1 CONVEXITY—DEFINITIONS AND PROPERTIES

No attempt is made to prove most of the basic properties of convex functions and convex sets stated in this section, nor is there any attempt at completeness. Convexity has been dealt with in detail in many places, and the reader is referred to [118], [10], or [68] for fuller development.

Basic to the notion of convexity is that of a convex set of points. We confine ourselves to the properties most essential to the ensuing discussion.

Definition. *A set of points T is a convex set if every convex combination of points in T is also in T.* That is, for every λ, $0 \leq \lambda \leq 1$, and any two points $x_1, x_2 \in T$, $[\lambda x_1 + (1 - \lambda)x_2] \in T$.

An important fact about convex sets is stated in the following lemma.

Lemma 8. The intersection of a finite number of convex sets is a convex set. That is, if each S_i ($i = 1, \ldots, m$) is a convex set, then

$$S = \bigcap_{i=1}^{m} S_i \text{ is a convex set.}$$

Proof. If S contains one or no points the lemma is true. If x_1 and x_2 are two points in S, then by the properties of intersection $x_1 \in S_i$ ($i = 1, \ldots, m$), $x_2 \in S_i$ ($i = 1, \ldots, m$). Then, also, $[\lambda x_1 + (1 - \lambda)x_2] \in S_i$ ($i = 1, \ldots, m$), and hence $[\lambda x_1 + (1 - \lambda)x_2] \in \bigcap_{i=1}^{m} S_i$. Q.E.D.

The second important notion is convexity of a function.

Definition. *A function f is a convex function of x in a nonempty convex set S if for every two points $x_1 \in S$ and $x_2 \in S$, and every λ, where $0 \leq \lambda \leq 1$,*

$$f[\lambda x_1 + (1 - \lambda)x_2] \leq \lambda f(x_1) + (1 - \lambda)f(x_2). \tag{6.1}$$

The function f is *strictly* convex if strict inequality holds in (6.1) when $0 < \lambda < 1$ and $x_1 \neq x_2$.

If the convex set S in which the function f is convex consists of the entire space, we simply say that f is a convex function of x, or f is *convex*.

88 Convex Programming

If f is continuously differentiable for $x \in S$, then an equivalent definition of a convex function is that

$$f(x_2) \geq f(x_1) + (x_2 - x_1)^T \nabla f(x_1) \tag{6.2}$$

for all $x_1, x_2 \in S$. Then f is *strictly convex* in S if strict inequality holds in (6.2) whenever $x_1 \neq x_2$.

Finally, if $f(x)$ is twice continuously differentiable, another way to define a convex function is that

$$\nabla^2 f(x) \text{ is a positive semidefinite matrix for all } x \in S. \tag{6.3}$$

(The definition of a positive semidefinite matrix is given in Section 2.2, Corollary 5.) If $\nabla^2 f(x)$ is positive definite, then $f(x)$ is strictly convex.

The two notions, convexity of a function and convexity of a set, are connected in the following lemma.

Lemma 9. *If $f(x)$ is a convex function in a convex set S, then, for any k, $S_k = \{x \mid f(x) \leq k, x \in S\}$ is a convex set (possibly empty).*

Proof. If S_k contains one or no points the theorem is true. If x_1 and x_2 are two points in S_k, then $\lambda x_1 + (1 - \lambda)x_2 \in S$, and $f[\lambda x_1 + (1 - \lambda)x_2] \leq \lambda f(x_1) + (1 - \lambda)f(x_2) \leq \lambda k + (1 - \lambda)k = k$. Thus $[\lambda x_1 + (1 - \lambda)x_2] \in S_k$.
Q.E.D.

Another useful property of convex functions is their additivity.

Lemma 10. *If $f_1, \ldots, f_p$ are convex functions in a convex set S, then $f(x) = \sum_{i=1}^{p} f_i(x)$ is a convex function in S.*

Proof. If x_1 and x_2 are two points in S,

$$f[\lambda x_1 + (1 - \lambda)x_2] = \sum_{i=1}^{p} f_i[\lambda x_1 + (1 - \lambda)x_2]$$

$$\leq \sum_{i=1}^{p} [\lambda f_i(x_1) + (1 - \lambda)f_i(x_2)]$$

$$= \lambda \sum_{i=1}^{p} f_i(x_1) + (1 - \lambda) \sum_{i=1}^{p} f_i(x_2) = \lambda f(x_1) + (1 - \lambda)f(x_2).$$
Q.E.D.

Definition. *A function $g(x)$ is* concave *in a convex set S if $-g(x)$ is a convex function in S.*

It is worth pointing out several facts that follow from the definition of concave and the lemmas proved for convex functions.

If $g(x)$ is a concave function in a convex set S, then

(i) $$g[\lambda x_1 + (1 - \lambda)x_2] \geq \lambda g(x_1) + (1 - \lambda)g(x_2), \tag{6.4}$$

where $0 \leq \lambda \leq 1$, $x_1, x_2 \in S$;

(ii) $$g(x_1) \leq g(x_2) + (x_1 - x_2)^T \nabla g(x_2) \qquad (6.5)$$

for $x_1, x_2 \in S$, when g is continuously differentiable;

(iii) $$\nabla^2 g(x) \text{ is a negative semidefinite matrix} \qquad (6.6)$$

for all $x \in S$ when g is twice continuously differentiable [that is, $z^T \nabla^2 g(x) z \leq 0$ for all z];

(iv) $$T_k = \{x \mid g(x) \geq k, x \in S\} \text{ is a convex set} \qquad (6.7)$$

(possibly empty) for all k; and
(v) if g_i, $i = 1, \ldots, p$ are concave functions in S, then

$$g(x) = \sum_{i=1}^{p} g_i(x) \text{ is a concave function in } S. \qquad (6.8)$$

6.2 CONVEX PROGRAMMING—THEORY

Definition. The convex programming *problem is written*

$$\text{minimize } f(x) \qquad (C)$$

subject to
$$g_i(x) \geq 0, \quad i = 1, \ldots, m, \qquad (6.9)$$

where $f(x)$, $\{-g_i(x)\}$ are convex continuous functions.

The remarkable property that every local minimum of (C) is a global minimum is proved in the following theorem.

Theorem 18 [Local-Global Convexity Property]. Every local minimum x^* of the convex programming problem (C) is a global minimum.

Proof. It follows from the concavity of the g_i's and the properties and lemmas of the previous section that $R = \{x \mid g_i(x) \geq 0, i = 1, \ldots, m\}$ is a convex set and that if x_1 and x_2 satisfy (6.9), then $\lambda x_1 + (1 - \lambda) x_2$ satisfies (6.9) for all $0 \leq \lambda \leq 1$.

Since x^* is a local minimum, it follows from the definition of a local minimum that there is a compact set V, where x^* is in the interior of V and

$$f(x^*) = v^* = \min_{R \cap V} f(x).$$

Let x^0 be any other point in R but not necessarily in V. Let λ be such that $\lambda x^* + (1-\lambda)x^0 \in R \cap V$, and $0 < \lambda < 1$. Then

$$f(x^0) \geq \frac{f[\lambda x^* + (1-\lambda)x^0] - \lambda f(x^*)}{1-\lambda} \quad \text{(convexity of } f\text{)}$$

$$\geq \frac{f(x^*) - \lambda f(x^*)}{1-\lambda} \quad (x^* \text{ is minimum in } R \cap V)$$

$$= f(x^*). \qquad \text{Q.E.D.}$$

For the convex programming problem, when the problem functions are differentiable, the Kuhn-Tucker first-order necessary conditions (Corollary 1) are also sufficient that x^* be a local (and, by Theorem 18, a global) constrained minimum of Problem C. That result is summarized in the following theorem.

Theorem 19 [Kuhn-Tucker Sufficiency Theorem]. If the functions f and $\{g_i\}$ are continuously differentiable, then sufficient conditions that x^* be a solution of the convex programming problem (C) is that there exist scalars $u_1^*, \ldots, u_m^*$ such that

$$g_i(x^*) \geq 0, \quad i = 1, \ldots, m, \qquad (6.10)$$

$$u_i^* g_i(x^*) = 0, \quad i = 1, \ldots, m, \qquad (6.11)$$

$$u_i^* \geq 0, \quad i = 1, \ldots, m, \qquad (6.12)$$

$$\nabla f(x^*) - \sum_{i=1}^m u_i^* \nabla g_i(x^*) = 0. \qquad (6.13)$$

Proof. Let x^0 be any other point satisfying the constraints (6.9). Then

$$f(x^0) \geq f(x^0) - \sum_{i=1}^m u_i^* g_i(x^0) \quad [(6.9), (6.12)]$$

$$\geq f(x^*) - \sum_{i=1}^m u_i^* g_i(x^*) + (x_0 - x^*)^T \left[\nabla f^* - \sum_{i=1}^m u_i^* \nabla g_i^* \right]$$

(convexity of f, concavity of g_i's)

$$= f(x^*) \quad [(6.11), (6.13)]. \qquad \text{Q.E.D.}$$

Since, when the first-order constraint qualification is assumed (Section 2.1), (6.10–6.13) are also necessary conditions, Corollary 1 and Theorem 19 thus constitute necessary and sufficient conditions for x^* to be a constrained minimum of Problem C.

6.2 Convex Programming—Theory

It is of interest that the second-order necessary conditions (2.13) are satisfied automatically here because the matrix $\nabla^2 f^* - \sum_{i=1}^{m} u_i^* \nabla^2 g_i^*$ is positive semidefinite when (assuming twice differentiability) f is convex and the g_i's are concave (see Section 6.1).

If the problem functions are twice differentiable Theorem 19 would follow directly from the neighborhood sufficiency theorem (Theorem 6, Section 2.3). The difficulties avoided by the assumption of convexity are apparent when the general theorems of Sections 2.1, 2.2, and 2.3 are compared with those of this chapter.

The analog of the sufficiency Theorem 19 can be stated for nondifferentiable functions.

Theorem 20 [Nondifferentiable Sufficiency Theorem]. Sufficient conditions that a point x^* be a solution to the convex programming problem (C) are that there exist scalars $u_1^*, \ldots, u_m^*$ such that (6.10–6.12) are satisfied and

$$\mathcal{L}(x^*, u^*) = \inf_x \left[\mathcal{L}(x, u^*) = f(x) - \sum_{i=1}^{m} u_i^* g_i(x) \right]. \tag{6.14}$$

Proof. The proof is actually simpler than that for Theorem 19. Basically (6.13) expresses the necessary and sufficient condition that x^* be an unconstrained minimum to the convex function $\mathcal{L}(x, u^*)$. Thus its choice for a replacement when the functions are not differentiable is obviously (6.14).

Let x^0 be any other point in R. Then

$$f(x^0) \geq f(x^0) - \sum_{i=1}^{m} u_i^* g_i(x^0) = \mathcal{L}(x^0, u^*)$$

$$\geq \inf_x \mathcal{L}(x, u^*) = \mathcal{L}(x^*, u^*)$$

$$= f(x^*). \qquad \text{Q.E.D.}$$

In order to establish a corresponding converse theorem we must replace the first-order constraint qualification (which depends on the differentiability of the functions) with a similar statement. One choice, which is stronger (for the convex programming problem) than the first-order constraint qualification when the functions are differentiable, is the requirement that there exist feasible points interior to the nonlinear constraints of (6.9). Then we have the following theorem for an isolated local minimum.

Theorem 21 [Nondifferentiable Necessary Conditions]. Necessary conditions that a point x^* be a solution to the convex programming problem (C) when there exist feasible points interior to the nonlinear constraints of (6.9) are that there exist scalars $u_1^*, \ldots, u_m^*$ satisfying (6.10–6.12) and (6.14).

Proof. The proof will be deferred to Section 6.3, where it will be a by-product of the way in which the exterior point unconstrained minimization algorithm approaches the solution of convex programming problems.

Duality Theory

For convex programming a body of theory has been developed in recent years on the subject of dual programming problems. This work is based on the fact that (6.13) which establishes one of the conditions that x^* be a solution, also states that x^* is the unconstrained minimum of the convex Lagrangian function $\mathcal{L}(x, u^*)$. As stated before, this is a property only of convex programming problems. In Section 2.2, the second-order conditions show that for nonconvex problems the situation is much more complicated. We call Problem C the *primal convex programming problem*. Using just (6.12) and (6.13), and ignoring the primal feasibility requirement (6.10) and the complementarity condition (6.11), the basic differentiable form of the dual programming problem [111] is to find values $[x^D, u^D]$ that solve the problem

$$\text{maximize } \mathcal{L}(x, u) \tag{D}$$

subject to

$$\nabla_x \mathcal{L}(x, u) = 0, \tag{6.15}$$

$$u_i \geq 0, \, i = 1, \ldots, m. \tag{6.16}$$

This problem is "dual" to (C) in the sense stated in the following two theorems.

Theorem 22 [Primal-Dual Bounds]. If f and $-\{g_i\}$ are convex functions, $y \in R$ is any primal feasible point, and (x^0, u^0) is any dual feasible point, then

$$f(y) \geq \mathcal{L}(x^0, u^0).$$

Proof.

$$f(y) \geq f(y) - \sum_{i=1}^{m} u_i^0 g_i(y)$$

$$\geq f(x^0) - \sum_{i=1}^{m} u_i^0 g_i(x^0) + (y - x^0)^T [\nabla f(x^0) - \sum u_i^0 \nabla g_i(x^0)]$$

$$= f(x^0) - \sum_{i=1}^{m} u_i^0 g_i(x^0) = \mathcal{L}(x^0, u^0). \qquad \text{Q.E.D.}$$

A nondifferentiable version of the duality theorem is obtained by replacing (6.15) with

$$\mathcal{L}(x^0, u^0) = \inf_x \mathcal{L}(x, u^0). \tag{6.17}$$

The proof of Theorem 22 is even more obvious, with (6.17) replacing (6.15).

6.2 Convex Programming—Theory 93

One of the practical values of Theorem 22 is that if dual feasible points are generated by algorithms that solve (C), then a lower bound is available on v^*, the optimum value of (C).

The connection between the conditions of Theorem 19, which are necessary when the first-order constraint qualification is assumed, and the duality theorems is apparent.

Theorem 23 [Existence of Solution to Dual]. If the first-order constraint qualification holds at x^*, a solution to Problem C, then there exists a solution to Problem D, and the maximum value of $\mathcal{L}$ is equal to the minimum value of f for $x \in R$.

Proof. Let u^* be the vector whose existence at x^* is guaranteed by Corollary 1. By Theorem 22, $\mathcal{L}(x^*, u^*) \leq v^*$. Using (2.7),

$$\mathcal{L}(x^*, u^*) = v^*,$$

and thus (x^*, u^*) must be a solution of (D). Q.E.D.

In certain cases it is possible to modify (D) and eliminate dependence on x. Problem D can be stated solely in terms of a maximization problem in u, giving it a true "dual" flavor. A development of this and an unconstrained minimization algorithm resulting from it are contained in Section 7.4.

Perturbation of Compact Convex Sets

A final property of convexity has to do with the preservation of boundedness properties of convex sets given by systems of inequalities. This property is a global generalization of the remarks following Theorem 7, indicating that for a perturbation of the inequalities defining an isolated bounded set of local minima, a bounded set resulted. The next theorem is used eventually to show that certain interior point and exterior point unconstrained minimization functions have finite minima for *all* values of their respective parameters.

Theorem 24 [Preservation of Boundedness of Convex Sets Given by Concave Inequalities]. If g_i ($i = 0, 1, \ldots, m$) are concave functions, and if $G = \{x \mid g_i(x) \geq 0, i = 0, 1, \ldots, m\}$ is a nonempty bounded set, then for any set of values $\{\epsilon_i\}$, where $\epsilon_i \geq 0, i = 0, 1, \ldots, m$, the set

$$\{x \mid g_i(x) \geq -\epsilon_i, i = 0, 1, \ldots, m\}$$

is bounded.

Proof. Obviously it suffices to show that when $\epsilon_0 > 0$ and $\epsilon_i = 0$, $i = 1, \ldots, m$, $G_\epsilon = \{x \mid g_0(x) \geq -\epsilon_0, g_i(x) \geq 0, i = 1, \ldots, m\}$ is bounded.

Assume the contrary; then, from any $x_1 \in G$ [G is nonempty and convex by hypothesis], there exists a ray that intersects the boundary of G but not of

G_ϵ. Let x_2 be a point on that ray such that $g_0(x_2) = -\delta < 0$, $g_i(x_2) \geq 0$, $i = 1, \ldots, m$. From the concavity of g_0,

$$-\delta = g_0(x_2) \geq \lambda g_0\left\{x_1 + \frac{1}{\lambda}(x_2 - x_1)\right\} + (1 - \lambda)g_0(x_1),$$

where $0 < \lambda < 1$. Rewriting yields

$$g_0\left\{x_1 + \frac{1}{\lambda}(x_2 - x_1)\right\} \leq \frac{-\delta - (1 - \lambda)g_0(x_1)}{\lambda} \leq \frac{-\delta}{\lambda}$$

since $g_0(x_1) \geq 0$ and $1 - \lambda > 0$. Choosing $\lambda < \delta/\epsilon_0$ shows that there is a point on the ray from x_1 passing through x_2 that takes on values smaller than $-\epsilon_0$. Q.E.D.

Corollary 20. If $f, -g_1, \ldots, -g_m$ are convex functions, and if the set of minima of Problem C is a nonempty bounded (and hence compact) set, then for every finite $k_0, k_1, \ldots, k_m$ the set

$$S_k = \{x \mid f(x) \leq k_0, g_i(x) \geq k_i, i = 1, \ldots, m\}$$

is a compact set (possibly empty).

Proof. Make the correspondence

$$g_0(x) = -f(x) + v^*,$$

where $v^* = \inf_R f(x)$, and the corollary follows from Theorem 24. Q.E.D.

6.3 SOLUTION OF CONVEX PROGRAMMING PROBLEMS BY INTERIOR POINT UNCONSTRAINED MINIMIZATION ALGORITHMS

We place restrictions on the function I so that the convexity of the g_i's can be used to advantage.

Lemma 11 [Convexity of the Unconstrained Function]. If I is a *convex decreasing* function of $g = (g_1, \ldots, g_m)^T$ for $g > 0$, where each g_i is a concave function, then $I[g(x)]$ is a convex function of x in $R^0 = \{x \mid g(x) > 0\}$. Furthermore, if f is a convex function of x, then $U(x, r_k) = f(x) + r_k I[g(x)]$ is a convex function of x in R^0 when $r_k > 0$.

Proof. Suppose x_1 and $x_2 \in R^0$, then $(0 \leq \lambda \leq 1)$

$$g[\lambda x_1 + (1 - \lambda)x_2] \geq \lambda g(x_1) + (1 - \lambda)g(x_2) > 0$$

(concavity of each component of g). Hence

$$I\{g[\lambda x_1 + (1 - \lambda)x_2]\} \leq I[\lambda g(x_1) + (1 - \lambda)g(x_2)]$$

6.3 Solution of Convex Problems by Interior Point Algorithms

(the decreasing property of I in g)

$$\leq \lambda I[g(x_1)] + (1 - \lambda)I[g(x_2)] \quad (6.18)$$

(the convexity of I in g).

The second part of the lemma follows from the convexity properties of Section 6.1 and (6.18). Q.E.D.

Note that the functions $\sum_{i=1}^{m} 1/g_i$ and $-\sum_{i=1}^{m} \ln g_i$ both are defined, decreasing, and convex in g when $g > 0$.

Another important question is: Under what conditions does the function $U(x, r_k)$ have an unconstrained minimum at a finite point for *every* $r_k > 0$ in a convex programming problem? From Theorem 8 we know that when r_k is small (and > 0) a finite minimum exists if the set of minimizing x's is compact. Clearly, if R, the set of feasible points, is bounded, a minimum of $U(x, r_k)$ exists for *every* $r_k > 0$. We shall prove that for the convex programming problem, if the set of solution points is bounded and if $I(g)$ has some additional mild restrictions placed on it, a finite minimum exists for every $r_k > 0$.

Lemma 12 [Boundedness of U Contours]. If the set of points that solve the convex programming problem (C) is bounded (hence compact), R^0 is nonempty, and $I(g)$ has the properties that (a) $\lim_{l \to \infty} I(g^l) = +\infty$ for every infinite sequence $\{g^l\}$, where $g^l > 0$ for all l and $\lim_{l \to \infty} g_i^l = 0$ for some i, (b) I is defined and decreasing in g when $g > 0$, and (c) whenever $g^0 > 0$, $d^0 \geq 0$, and $\{\lambda^l\}$ is a strictly positive null sequence,

$$\lim_{l \to \infty} \lambda^l I\left(g^0 + \frac{d^0}{\lambda^l}\right) = 0, \quad (6.19)$$

then in R^0 for any $r_k > 0$, the isocontours of $U(x, r_k) = f(x) + r_k I[g(x)]$ are bounded. If R, the set of feasible points, is bounded, the conclusion follows whether or not (6.19) holds for I.

Proof. The last part of the theorem is trivial, since $I(g)$ is infinite at every boundary point of R and $I(g)$ is continuous in R^0.

To prove the first part we assume the contrary, that for some $r_k > 0$, $\{y^l\}$ is an unbounded sequence of interior feasible points for which $U(y^l, r_k)$ is bounded above. Let x^0 be any point in R^0 and let $D \equiv \{x \mid f(x) \leq f(x^0) + \epsilon, x \in R\}$ for some $\epsilon > 0$. By our assumption $\{x \mid f(x) \leq v^* = \inf_R f(x), x \in R\}$ is nonempty and bounded. By Theorem 24, then D is also bounded, and hence compact. Let B denote the boundary of D. (Note that B is compact also.)

Consider the point y^l far enough out in the sequence so that $y^l \notin D$.

96   Convex Programming

Let z^l be the point on the boundary B of D where the line connecting x^0 and y^l intersects B. By definition of D, $z^l \neq x^0$. Let λ^l be the scalar where, for $0 < \lambda^l < 1$,
$$z^l = \lambda^l y^l + (1 - \lambda^l) x^0.$$
Clearly $\lambda^l \to 0$ as $l \to \infty$. By concavity,
$$g_i(z^l) \geq \lambda^l g_i(y^l) + (1 - \lambda^l) g_i(x^0) > 0, \qquad i = 1, \ldots, m. \quad (6.20)$$
Hence $z^l \in R^0$, but since z^l is a boundary point of D, it must lie on that portion of the boundary $B \subset R^0$ and such that
$$f(z^l) = f(x^0) + \epsilon. \quad (6.21)$$
By convexity,
$$f(y^l) \geq \frac{f(z^l) - (1 - \lambda^l) f(x^0)}{\lambda^l}$$
$$= f(x^0) + \frac{\epsilon}{\lambda^l}. \quad (6.22)$$
Since $\epsilon > 0$, $f(y^l) \to +\infty$ as $\lambda^l \to 0$ $(l \to \infty)$.

Rewriting (6.20), and using the fact that $y^l \in R^0$,
$$0 < g_i(y^l) \leq \frac{g_i(z^l) - (1 - \lambda^l) g_i(x^0)}{\lambda^l}$$
$$\leq \max_{x \in B} g_i(x^0) + \frac{g_i(x) - g_i(x^0)}{\lambda^l},$$
$$= g_i(x^0) + \frac{d_i^0}{\lambda^l}, \qquad i = 1, \ldots, m, \quad (6.23)$$
where d_i^0 denotes $\max_{x \in B} [g_i(x) - g_i(x^0)]$, $i = 1, \ldots, m$. (This exists because B is compact.) Let $d^0 = (d_1^0, \ldots, d_m^0)^T$. That $d_i^0 > 0$, each i, follows from (6.23); otherwise the right-hand term of (6.23) would become negative as λ^l went to zero.

Then, using (6.22), (6.23), and the decreasing property of I,
$$U(y^l, r_k) = f(y^l) + r_k I[g(y^l)]$$
$$\geq f(x^0) + \frac{\epsilon}{\lambda^l} + r_k I\left[g(x^0) + \frac{d^0}{\lambda^l} \right]$$
$$= f(x^0) + \frac{\epsilon + r_k \lambda^l I[g(x^0) + d^0/\lambda^l]}{\lambda^l}. \quad (6.24)$$

By our assumption (Property c on I) the second term in the numerator above goes to 0 as $\lambda^l \to 0$. Thus $\lim_{l \to \infty} U(y^l, r_k) = +\infty$, contradicting our initial statement.   Q.E.D.

Note that no convexity assumption was made on I.

6.3 Solution of Convex Problems by Interior Point Algorithms

It is a trivial corollary of this lemma that all minima of $U(x, r_k)$ in R^0 must be taken on at finite points.

It is easy to verify that the logarithmic penalty function $I(g) = -\sum_{i=1}^{m} \ln g_i$ satisfies (6.19). Taking the limit of the negative of any term and letting $d_i^0 > 0$, $g_i^0 > 0$,

$$0 \leq \lim_{l \to \infty} \lambda^l \ln \left[g_i^0 + \frac{d_i^0}{\lambda^l} \right]$$

$$\leq \lim_{l \to \infty} \lambda^l \ln \left[\frac{g_i^0 + d_i^0}{\lambda^l} \right]$$

$$= \lim_{l \to \infty} \lambda^l \ln [g_i^0 + d_i^0] - \lambda^l \ln \lambda^l$$

$$= 0.$$

The function $I_i(g_i) = 1/g_i$ clearly satisfies (6.19).

If for some $I(g)$ (6.19) does not hold, the following example demonstrates the possibility that a finite unconstrained minimum may not exist for every $r_k > 0$.

Example. Suppose that the programming problem is

$$\text{minimize } x_1$$

subject to

$$g_1(x) \equiv x_1 \geq 0.$$

Let $I_1(g_1) = -\ln g_1$ when $0 < g_1 \leq 1$, and $I_1(g_1) = -g_1 + 1$ when $1 < g_1$. Then $U(x, r_k) = x_1 + r_k I_1[g_1(x_1)]$ has an unconstrained minimum for $r_k < 1$ at $x(r_k) = r_k$; for $r_k = 1$ it is minimized at any point $x_1 \geq 1$ and for $r_k > 1$ it is minimized at $x_1 = +\infty$. In this example all the hypotheses of the theorem are satisfied except that (6.19) does not hold.

We are now in a position to prove our basic convergence theorem for convex programming.

Theorem 25 [Convergence to Solution of Primal Convex Programming Problem by Interior Point Algorithms]. If f and $-\{g_i\}(i = 1, \ldots, m)$ are convex functions; the set of points that solve the primal convex programming problem (C) is bounded (hence compact); the set

$$R^0 = \{x \mid g_i(x) > 0, i = 1, \ldots, m\}$$

is nonempty; the function $I[g(x)]$ has the properties that (a) $I(g)$ is *defined*, *decreasing*, and *convex* in g when $g > 0$, and $\lim_{l \to \infty} \lambda^l I[g^0 + d^0/\lambda^l] = 0$ whenever $g^0 > 0$, $d^0 \geq 0$, for any strictly positive null sequence $\{\lambda^l\}$, and

(b) for any strictly positive sequence $\{g^l\}$, $\lim_{l \to \infty} I(g^l) = +\infty$, if $\lim_{l \to \infty} g_i^l = 0$ for some i; and $\{r_k\}$ is a strictly decreasing null sequence, then

(i) the interior point unconstrained minimization function $U(x, r_k) = f(x) + r_k I[g(x)]$ has a finite unconstrained minimum $x(r_k) \in R^0$ for every $r_k > 0$,

(ii) $U(x, r_k)$ is a convex function in R^0,

(iii) every local unconstrained minimum of $U(x, r_k)$ in R^0 is a *global* unconstrained minimum of $U(x, r_k)$ in R^0,

(iv) $\lim_{k \to \infty} f[x(r_k)] = \lim_{k \to \infty} U[x(r_k), r_k] = v^* = \min_R f(x)$,

(v) every limit point of $\{x(r_k)\}$ solves the convex programming problem (C),

(vi) $\{U[x(r_k), r_k]\}$ is a monotonically decreasing sequence if $I(g) > 0$ for $g > 0$,

(vii) $\{f[x(r_k)]\}$ is a monotonically decreasing sequence, and

(viii) $\{I[x(r_k)]\}$ is a monotonically increasing sequence.

Proof. Part i follows from Lemma 12, Part ii is Lemma 11, Part iii follows from the convexity of U, and Parts iv and v follow from Theorem 8.

Let $f[x(r_k)] = f^k$, and so on. Because each x^k is a global minimum of U in R^0,

$$f^k + r_k I^k \leq f^{k+1} + r_k I^{k+1}, \tag{6.25}$$

$$f^{k+1} + r_{k+1} I^{k+1} \leq f^k + r_{k+1} I^k.$$

Multiplying (6.25) by r_{k+1}/r_k, adding to the second inequality, and rearranging yields

$$\left(1 - \frac{r_{k+1}}{r_k}\right) f^{k+1} \leq \left(1 - \frac{r_{k+1}}{r_k}\right) f^k.$$

Since $r_{k+1} < r_k$, Part vii follows. By (vii) and (6.25), (viii) follows. Q.E.D.

Solution of Dual Convex Programming Problem

When $I(g)$ is defined and differentiable for each $g_i > 0$ [for example, $\partial(-\sum_{j=1}^m \ln g_j)/\partial g_i = -1/g_i$], it is possible to show that interior point unconstrained minimization algorithms yield points that are feasible to the nondifferentiable form of the dual convex programming problem (D) at every iteration and in the limit solve this problem.

In the following it is assumed that x^* is one of the limit points of $\{x(r_k)\}$.

Theorem 26 [Dual Convergence–Interior Point Algorithms]. If in addition to the assumptions of Theorem 25 $I(g)$ is separable in g [that is, $I(g) = \sum_{i=1}^m I_i(g_i)$] and each $I_i(g_i)$ is differentiable in g_i when $g_i > 0$, then

$$[x(r_k), u(r_k)]$$

6.3 Solution of Convex Problems by Interior Point Algorithms

is a point feasible to the constraints (6.17) and (6.16) of the dual problem (D), where

$$u_i(r_k) = -r_k \frac{\partial I_i\{g_i[x(r_k)]\}}{\partial g_i}, \quad i = 1, \ldots, m, \tag{6.26}$$

all limit points of this sequence are finite solutions of (D), and at least one limit point exists.

Proof. Since each I_i is decreasing in g_i, (6.26) satisfies (6.16). For simplicity of notation let x^k denote $x(r_k)$, and so on. Let y be any point in R. By convexity,

$$f(y) - f(x^k) \geq \theta^{-1}[f(z) - f(x^k)], \tag{6.27}$$

where $z = x^k + \theta(y - x^k)$, $0 < \theta < 1$.

Because x^k minimizes $U(x, r_k)$ in R^0, and because

$$g_i(z) \geq (1 - \theta)g_i(x^k) + \theta g_i(y) > 0, \quad i = 1, \ldots, m, \tag{6.28}$$

means $z \in R^0$, it follows that

$$U(z, r_k) \geq U(x^k, r_k). \tag{6.29}$$

Substituting (6.29) in (6.27) yields

$$f(y) - f(x^k) \geq \theta^{-1} r_k \{I[g(x^k)] - I[g(z)]\}$$

$$\geq \theta^{-1} r_k \sum_{i=1}^{m} [g_i(x^k) - g_i(z)] \frac{\partial I_i [g_i(z)]}{\partial g_i}$$

because I is convex and differentiable in each g_i. Since I_i is decreasing in g_i, $\partial I_i[g(z)]/\partial g_i \leq 0$. Using this and (6.28), substituting in the above yields

$$f(y) - f(x^k) \geq r_k \sum_{i=1}^{m} [g_i(x^k) - g_i(y)] \frac{\partial I_i [g_i(z)]}{\partial g_i}. \tag{6.30}$$

Taking the limit as $\theta \to 0$ yields $\mathcal{L}(y, u^k) \geq \mathcal{L}(x^k, u^k)$ for all $y \in R$. Because $x^k \in R^0$ and because $\mathcal{L}(x, u^k)$ is everywhere convex, then x^k is a global unconstrained minimum of $\mathcal{L}(x, u^k)$. Thus (6.17) is satisfied.

Let y^0 be any point in R^0. Substituting in (6.30) and rearranging yields

$$f(y^0) - \sum u_i^k g_i(y^0) \geq f(x^k) - \sum u_i^k g_i(x^k).$$

Clearly, because of the continuity of $\partial I_i/\partial g_i$ in g_i when $g_i > 0$ using (6.26), $\lim_{k \to \infty} u_i^k = 0$ for all $i \notin B^*$, that is, for all i, where $g_i(x^*) > 0$. Define $v^k = \sum_{i=1}^{m} u_i^k$, and divide both sides of (6.30) by v^k. If $\liminf_{k \to \infty} v^k = +\infty$, then there exists a set of values $\alpha_i \geq 0$, $\sum_{i \in B^*} \alpha_i = 1$, such that

$$-\sum_{i \in B^*} \alpha_i g_i(y^0) \geq 0.$$

Since $y^0 \in R^0$, this means $0 > 0$, a contradiction. Hence every limit point of $\{u^k\}$ is finite. Thus, $\lim_{k \to \infty} \mathcal{L}(x^k, u^k) = v^*$. Q.E.D.

In particular, the way in which an interior point algorithm converges, its dual properties, and resulting bounds are demonstrated in the following example.

Example. The convex programming problem in E^2 is

$$\text{minimize } [f(x) = -\min(x_1, x_2)]$$

subject to

$$g_1 = -x_1^2 - x_2^2 + 1 \geq 0.$$

The isocontours of the objective function and the constraint region are pictured in Figure 7. The solution is at $(\sqrt{2}/2, \sqrt{2}/2) = x^*$, with $f(x^*) = -\sqrt{2}/2 = v^*$. For our interior unconstrained minimization function we choose

$$L(x, r) = -\min(x_1, x_2) - r \ln(-x_1^2 - x_2^2 + 1).$$

It is clear from the geometric depiction that at the minimum of L for any $r > 0$, $x_1(r) = x_2(r)$. Substituting this in L yields

$$L(x, r) = -x_1 - r \ln(-2x_1^2 + 1).$$

Differentiating with respect to x_1, setting that equal to zero, and solving for $x_1(r)$ gives

$$x_1(r) = \frac{-2r \pm \sqrt{4r^2 + 2}}{2}.$$

Since $x_1(r) = x_2(r)$, and the minimizing point must be interior, the positive root must be taken. The dual variable given by (6.26) is then

$$u_1(r) = (-4r + \sqrt{16r^2 + 8})^{-1}.$$

According to Theorem 26, $[x_1(r), x_2(r)]$ minimizes

$$\mathcal{L}[x, u(r)] = -\min(x_1, x_2) - (-4r + \sqrt{16r^2 + 8})^{-1}(-x_1^2 - x_2^2 + 1)$$

and $\mathcal{L}[x(r), r] \leq v^* = -\sqrt{2}/2$. To verify this we note that

$$-\min(x_1, x_2) = \frac{-x_1 - x_2 + |x_1 - x_2|}{2}.$$

Then for any $u^0 > 0$ the minimum of

$$\frac{-x_1 - x_2 + |x_1 - x_2|}{2} - u^0(-x_1^2 - x_2^2 + 1)$$

6.3 Solution of Convex Problems by Interior Point Algorithms

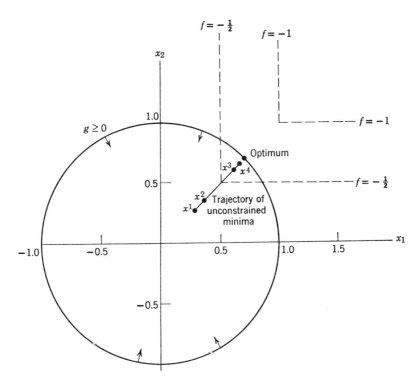

Figure 7 Solution of convex programming problem. The problem is:

$$\text{minimize} - \min(x_1, x_2)$$

subject to

$$-x_1^2 - x_2^2 + 1 \geq 0.$$

(since it is strictly convex and approaches $+\infty$ as $|x_1|, |x_2| \to +\infty$, and since it is symmetric in x_1 and x_2, it must be minimized at a finite point) is achieved at $x_1(u^0) = x_2(u^0) = \dfrac{1}{4u^0}$. Thus $x_1(r), x_2(r)$, and $u_1(r)$ satisfy this equality.

Furthermore

$$-\left(\frac{1}{4u^0}\right) - u^0\left[-2\left(\frac{1}{4u^0}\right)^2 + 1\right] = -\frac{1}{8u^0} - u^0.$$

This is a concave function in u^0 when $u^0 > 0$ and has a maximum at $u^0 = \sqrt{1/8}$, where it has the value $-\sqrt{2}/2$.

Note that $L(x, r)$ is not differentiable in x at $x_1(r) = x_2(r)$, nor is $\mathcal{L}$ differentiable at x^*. The dual variable, however, is defined and unique.

Several values of the minimizing x for different values of r and the corresponding dual values are given in Table 5. Note that they provide tighter and tighter bounds for the optimal objective function value.

We note one interesting fact about the logarithmic penalty function. Because of the dual property, it is possible to preset the difference between the primal and dual objective function values at the first r minimum, and thus

Table 5 Values of f, $\mathcal{L}$, and x for Example

r_k		$x_1{}^k$	$x_2{}^k$	f^k (Primal)	$\mathcal{L}^k$ (Dual)
r_1	1.000	0.225	0.225	-0.225	-1.225
r_2	0.500	0.366	0.366	-0.366	-0.866
r_3	0.100	0.614	0.614	-0.614	-0.714
r_4	0.010	0.697	0.697	-0.697	-0.707
Solution (approx.)		$+0.707$	0.707	-0.707	-0.707
Solution (theoret.)		$\sqrt{2}/2$	$\sqrt{2}/2$	$-\sqrt{2}/2$	$-\sqrt{2}/2$

in a sense solve the problem in one unconstrained minimization. Suppose that it is desired to know the optimum objective function value to within ϵ. Let $r_1 = \epsilon/m$. Then, since

$$f^1 \geq v^* \geq f^1 - \sum_{i=1}^{m} \frac{r_1}{g_i^1} g_i^1 = f^1 - \sum_{i=1}^{m} \frac{\epsilon}{m} = f^1 - \epsilon,$$

the optimum is known within ϵ. (f^1 denotes $f[x(r_1)]$, and so on.)

6.4 SOLUTION OF CONVEX PROGRAMMING PROBLEM BY EXTERIOR POINT UNCONSTRAINED MINIMIZATION ALGORITHMS

The development for exterior point algorithms parallels that in Section 6.3 for interior point methods. An analogous choice for the function $O[g(x)]$ must be made so that every $T(x, t_k) = f(x) + t_k O[g(x)]$ has a finite minimum. Except that R^0 may be empty, the same conditions of compactness and convexity are assumed as before.

Lemma 13 [Convexity of T Function]. If $O(g)$ is a *convex decreasing* function of the concave function g (each g_i a concave function) and if $O(g) = 0$ when $g_i \geq 0$ and $O(g) > 0$ when some $g_i < 0$, then $O[g(x)]$ is a convex function of x in E^n. Furthermore, if $f(x)$ is a convex function, then, for every $t_k > 0$, $T(x, t_k) = f(x) + t_k O[g(x)]$ is a convex function of x in E^n.

6.4 Solution of Convex Problems by Exterior Point Algorithms 103

Proof. The proof is analogous to that of Lemma 11. The only difference is that the set in which $O[g(x)]$ is convex is now E^n, not just R^0.

The function $T(x, t_k)$ has a finite minimum for every $t_k > 0$ if $O[g(x)] \to +\infty$ "faster" than $f(x) \to -\infty$ for any unbounded sequence of points and if the usual conditions on the set of minima to Problem C are assumed. This lemma corresponds to the portion of Theorem 9, where it was shown under more general conditions that for t_k large enough a finite minimum of $T(x, t_k)$ exists.

Lemma 14 [Existence of T Minima]. If f, $-g_i$, $i = 1, \ldots, m$ are convex functions, the set of minima of Problem C is compact, and $O(g_i)$ has the properties that (a) $O(g) = 0$ when $g \geq 0$ and (b) there exists a $\gamma > 0$ such that for every $\delta > 0$, whenever $g_i \leq -\delta$ for some i, it follows that

$$O(g) \geq \delta^{\gamma+1}, \quad . \tag{6.31}$$

then for every $t_k > 0$ there is a finite point that minimizes $T(x, t_k)$, and the set of points minimizing $T(x, t_k)$ is bounded.

Proof. We prove the equivalent assertion, that for every $t_k > 0$, and for every finite M,

$$S_M \equiv \{x \mid T(x, t_k) \equiv f(x) + t_k O[g(x)] \leq M\}$$

is a bounded set (possibly empty).

If $S_M = \emptyset$, the conclusion holds trivially. Assume that $S_M \neq \emptyset$ and the conclusion is false. Let $\{y^l\}$ be a sequence of points where $T(y^l, t_k) \leq M$ (for some M), and $\lim_{l \to \infty} |y^l| = +\infty$. Since $t_k O[g(y^l)] \geq 0$, $f(y^l)$ must be bounded above. But, by Corollary 20, $g_i(y^l) \to -\infty$ for some i.

By our assumption (6.31) $t_k O[g(y^l)] \to +\infty$ as $l \to \infty$, and therefore, for $T(y^l, t_k)$ to be bounded above, $f(x^l) \to -\infty$. We show that it cannot go to $-\infty$ fast enough to vitiate the desired conclusion.

Let $\epsilon > 0$ be any positive number, and

$$R(\epsilon) = \{x \mid f(x) \leq \max(M, v^*) + 1; g_i(x) \geq -\epsilon, i = 1, \ldots, m\}.$$

By Corollary 20, $R(\epsilon)$ is compact. It is not empty since x^* is contained in its interior $R^0(\epsilon)$. [Recall $f(x^*) = v^* = \inf_R f(x)$.] Consider points far out enough in the sequence so that $y^l \notin R(\epsilon), f(y^l) \leq v^*$.

Connect x^* and y^l. Let x_1 denote the point where the line joining x^* and y^l intersects the boundary of $R(\epsilon)$. Let λ^l, where $0 < \lambda^l < 1$, be the scalar such that $\lambda^l y^l + (1 - y^l)x^* = x_1$. [$\lambda^l \neq 0$ since $x^* \in R^0(\epsilon)$, and x_1 is on the boundary of $R(\epsilon)$]. Note that we must have $\lambda^l \to 0$ as $l \to \infty$, since $|y^l| \to +\infty$.

By convexity (and the selection of y^l),

$$f(x_1) \leq \lambda^l f(y^l) + (1 - \lambda^l) f(x^*) \leq v^*. \tag{6.32}$$

Then $f(x_1) < \max(M, v^*) + 1$, implying that x_1 is on the boundary portion of $R(\epsilon)$, where $g_i(x_1) = -\epsilon$ for at least one i. For convenience let this hold for $i = 1$; that is, $g_1(x_1) = -\epsilon$.

Let $v_1 = \min_{R(\epsilon)} f(x)$, which exists and is finite, since $R(\epsilon)$ is compact. Using this and the inequality obtained from (6.32) we have

$$f(y^l) \geq \frac{f(x_1) - (1 - \lambda^l)f(x^*)}{\lambda^l} \tag{6.33}$$

$$\geq \frac{v_1 - (1 - \lambda^l)v^*}{\lambda^l} = v^* + \frac{v_1 - v^*}{\lambda^l}.$$

From the concavity of each $g_i(x)$ and the fact that $g_i(x^*) \geq 0$, $i = 1, \ldots, m$, we have that

$$0 > -\epsilon = g_1(x_1) \geq \lambda^l g_1(y^l) + (1 - \lambda^l)g_1(x^*) \geq \lambda^l g_1(y^l).$$

Now from the property (6.31) assumed on $O(g)$,

$$t_k O[g(y^l)] \geq t_k \left(\frac{\epsilon}{\lambda^l}\right)^{\gamma+1} \quad \text{for some } \gamma > 0. \tag{6.34}$$

Summing (6.33) and (6.34) gives

$$v^* + \frac{v_1 - v^*}{\lambda^l} + t_k \left(\frac{\epsilon}{\lambda^l}\right)^{\gamma+1} \leq f(y^l) + t_k O[g(y^l)].$$

Clearly, since $t_k, \epsilon, \gamma > 0$, $\lim_{l \to \infty} T(y^l, t_k) = +\infty$. Q.E.D.

Note that for the quadratic loss function (6.31) is satisfied, and in particular $\gamma = 1$.

The basic convergence theorem for convex programming problems using exterior point algorithms is as follows.

Theorem 27 [Solution of Convex Problems by Exterior Point Algorithms]. If f, $-g_i$, $i = 1, \ldots, m$ are convex functions; the set of points solving (C) is compact; $O(g)$ has the properties that (a) O is *convex* and decreasing in g, (b) there exists a $\gamma > 0$ such that for every $\delta > 0$, where $g_i \leq -\delta < 0$, for some i, $O(g) \geq (\delta)^{\gamma+1}$, and (c) $O(g) = 0$ when $g \geq 0$; and $\{t_k\}$ is a positive, strictly increasing unbounded sequence, then

(i) For every $t_k > 0$, $T(x, t_k)$ is convex,
(ii) $T(x, t_k)$ is minimized at a finite point, x^k,
(iii) Every limit point of the uniformly bounded sequence $\{x^k\}$ solves the convex programming problem (C),
(iv) $\{T[x^k, t_k]\}$ is a monotonically increasing sequence, $T^k \to v^*$,
(v) $\{f^k\}$ is a monotonically increasing sequence, $f^k \to v^*$, and,
(vi) $\{O^k\}$ is a monotonically decreasing sequence.

6.4 Solution of Convex Problems by Exterior Point Algorithms

Proof. Parts i and ii follow from Lemmas 13 and 14. Because every $T(x, t_k)$ is convex, every local minimum is global; that is, if two points $x^1(t_k)$ and $x^2(t_k)$ minimize $T(x, t_k)$, $T[x^1(t_k), t_k] = T[x^2(t_k), t_k]$. Convergence of every limit point of the sequence to the compact set of minima of (C) is then guaranteed by Theorem 9. [Note that, since

$$\{x \mid T(x, t_k) \leq M\} \supset \{x \mid T(x, t_{k+1}) \leq M\}$$

when $t_k < t_{k+1}$, part iv holds and, also, all minimizing sequences are bounded.] Since x^k and x^{k+1} are global minima,

$$T[x^k, t_k] \leq T[x^{k+1}, t_k], \tag{6.35}$$

$$T[x^{k+1}, t_{k+1}] \leq T[x^k, t_{k+1}]. \tag{6.36}$$

Multiplying (6.35) by t_{k+1}/t_k, adding to (6.36), and rearranging yields

$$\left(\frac{t_{k+1}}{t_k} - 1\right)f^k \leq \left(\frac{t_{k+1}}{t_k} - 1\right)f^{k+1}.$$

Since $t_{k+1} > t_k$ by assumption, Part v follows and with (6.36) gives vi.
Q.E.D.

Exterior point algorithms also yield points that are dual feasible [satisfy (6.16) and (6.17)] and, in the limit, solve the dual programming problem.

Theorem 28 [Exterior Point Dual Convergence]. If in addition to the hypotheses of Theorem 27 $O(g)$ is separable and each $O_i(g_i)$ is continuously differentiable in g_i, then an exterior point algorithm yields points

$$[x(t_k), u(t_k)],$$

where

$$u_i(t_k) = -t_k \frac{\partial O_i\{g_i[x(t_k)]\}}{\partial g_i}, \quad i = 1, \ldots, m, \tag{6.37}$$

that are feasible to the dual convex programming problem (D) [satisfy (6.16) and (6.17)] and

$$\lim_{k \to \infty} \mathcal{L}[x(t_k), u(t_k)] = v^* = \inf_R f(x).$$

Proof. Note that (6.16) is satisfied, that is, $u^k \geq 0$, since $t_k > 0$ and $\partial O_i/\partial g_i \leq 0$. Let y be any point and let

$$z = x^k + \theta[y - x^k], \quad 0 < \theta < 1 \tag{6.38}$$

By convexity,

$$f(z) - f^k \leq \theta[f(y) - f^k], \tag{6.39}$$

and, by concavity,

$$g_i(z) \geq g_i^k + \theta[g_i(y) - g_i^k], \quad i = 1, \ldots, m. \tag{6.40}$$

It follows from (6.40) and the decreasing property of $O_i(g_i)$ that

$$O_i[g_i(z)] \leq O_i\{g_i^k + \theta[g_i(y) - g_i^k]\}. \tag{6.41}$$

Since x^k is the global minimum of $T(x, t_k)$,

$$f^k + t_k O^k \leq f(z) + t_k O[g(z)]. \tag{6.42}$$

Substituting (6.39) and (6.41) in (6.42) and rearranging,

$$t_k O^k \leq \theta[f(y) - f^k] + t_k \sum_{i=1}^{m} O_i\{g_i^k + \theta[g_i(y) - g_i^k]\}. \tag{6.43}$$

Using the assumed differentiability of each O_i in g_i and the convexity of O_i in g_i,

$$O_i(g_i^k) \geq O_i\{g_i^k + \theta[g_i(y) - g_i^k]\}$$

$$- \theta \sum_{i=1}^{m} [g_i(y) - g_i^k] \frac{\partial O_i\{g_i^k + \theta[g_i(y) - g_i^k]\}}{\partial g_i}. \tag{6.44}$$

Substituting (6.44) in (6.43), dividing by θ, and taking the limit as $\theta \to 0$ yields

$$0 \leq f(y) - f^k + \sum_{i=1}^{m} [g_i(y) - g_i^k] \left[\frac{\partial O_i^k}{\partial g_i}\right] t_k, \tag{6.45}$$

or $\mathcal{L}(y, u^k) \geq \mathcal{L}(x^k, u^k)$ for all y. Hence (6.17) is satisfied. Note that when $g_i[x(t_k)] \geq 0$, the corresponding u_i^k must be equal to zero, since the limit of $\partial O_i(g_i)/\partial g_i$ as $g_i \to 0^+$ is zero, and by the assumed continuity of the derivatives all directional derivatives are equal.

The dual bound is a better lower bound than $T(x^k, t_k)$. This is seen as follows. By the convexity of O_i in g_i,

$$t_k O_i(0) \geq t_k O_i(g_i^k) + t_k(0 - g_i^k)\left[\frac{\partial O_i(g_i^k)}{\partial g_i}\right],$$

or

$$g_i^k t_k \left[\frac{\partial O_i(g_i^k)}{\partial g_i}\right] \geq t_k O_i(g_i^k). \tag{6.46}$$

Summing over i and adding f^k to both sides,

$$\mathcal{L}(x^k, u_k) = f^k - \sum_{i=1}^{m}\left[-t_k \frac{\partial O_i^k}{\partial g_i}\right] g_i^k$$

$$\geq f^k + \sum_{i=1}^{m} t_k O_i(g_i^k) = T(x^k, t_k). \tag{6.47}$$

6.4 Solution of Convex Problems by Exterior Point Algorithms

Then, from Theorem 22 and (6.47),

$$v^* \geq \mathcal{L}(x^k, u^k) \geq T(x^k, t_k),$$

and taking the limit as $k \to \infty$ yields (see Part iv of Theorem 9) the final part of this theorem. Q.E.D.

Upper Bounds for Exterior Point Algorithms

At every minimum we have a lower bound on v^* given by the value of the dual objective function and the value of the exterior function itself. Another desirable feature would be an upper bound at every minimum that converged to v^* as $r_k \to 0$. If the interior of the feasible region (R^0) is nonempty, such a sequence of bounds can be generated. This result is stated in the next theorem.

Theorem 29 [Upper Bounds for v^*]. If $f(x)$, $-g_i(x)$, $i = 1, \ldots, m$ are convex functions, the set of points that solve the convex programming problem (C) is compact, and if $R^0 = \{x \mid g_i(x) > 0, i = 1, \ldots, m\}$ is nonempty, then, using the points $x(t_k)$ that minimize the T function, it is possible to generate a sequence $\{z^k\}$ such that $g_i(z^k) \geq 0$, $i = 1, \ldots, m$, where every limit point of $\{z^k\}$ is a limit point of $\{x^k\}$ (is an optimum).

Proof. Let x^k denote the point $x(t_k)$, and let y^0 be some interior point. (Such a point exists by hypothesis.) Then z^k is constructed as follows.

Let $I_k = \{i \mid g_i(x^k) = -\lambda_{i1}^k < 0\}$. Let $g_i(y^0) = \lambda_{i2}^0 > 0$, $i = 1, \ldots, m$. Calculate

$$\mu^k = \max_{i \in I_k} \frac{\lambda_{i1}^k}{\lambda_{i1}^k + \lambda_{i2}^0}. \tag{6.48}$$

Define

$$z^k = (1 - \mu^k)x^k + \mu^k y^0. \tag{6.49}$$

Because of the concavity of the g_i's,

$$g_i(z^k) \geq (1 - \mu^k)g_i(x^k) + \mu^k g_i(y^0)$$
$$= (1 - \mu^k)(-\lambda_{i1}^k) + \mu^k(\lambda_{i2}^0) \geq 0 \quad \text{for all} \quad i \in I_k.$$

[Clearly $g_i(z^k) > 0$ for all $i \notin I_k$.] Hence z^k is feasible. Hence $f(z^k)$ provides an upper bound for v^*, and

$$f(z^k) \geq v^* \geq \mathcal{L}(x^k, u^k). \tag{6.50}$$

Because of (6.48), the fact that $\lambda_{i1}^k \to 0$ for all $i \in I_k$ (x^k becomes feasible in the limit), and $y^0 \in R^0$, $\lim_{k \to \infty} \mu^k = 0$. Hence from (6.49) it follows that limit points of $\{z^k\}$ are limit points of $\{x^k\}$. Q.E.D.

The bounds on v^* given by (6.50) provide a convenient convergence criterion for the algorithm.

Before the dual convergence Theorem 28 can be used to establish Theorem 21, the following lemma is needed.

Lemma 15. If x^* is an unconstrained minimum of the function $\varphi(x) + \|x - x^*\|^2$, where $\varphi(x)$ is a convex function, then x^* is also a (global) unconstrained minimum of $\varphi(x)$.

Proof. Let y be any point. Denote $y - x^*$ by s. Since x^* minimizes $\varphi(x) + \|x - x^*\|^2$,

$$\varphi(x^* + \theta s) + \|x^* + \theta s - x^*\|^2 \geq \varphi(x^*) + \|x^* - x^*\|^2 \quad (6.51)$$

for all θ, in particular for $1 > \theta > 0$. Using the basic inequality for convex functions,

$$\varphi(y) - \varphi(x^*) \geq \frac{\varphi(x^* + \theta s) - \varphi(x^*)}{\theta}, \quad 1 > \theta > 0. \quad (6.52)$$

Rearranging (6.51) and dividing by θ yields

$$\frac{\varphi(x^* + \theta s) - \varphi(x^*)}{\theta} \geq -\theta \|s\|^2. \quad (6.53)$$

Combining (6.52) and (6.53) and taking the limit as $\theta \to 0$ yields

$$\varphi(y) - \varphi(x^*) \geq 0 \quad \text{for all} \quad y. \qquad \text{Q.E.D.}$$

We are now ready to prove Theorem 21. It is necessary to rewrite the constraints of (C) as

$$g_i(x) \geq 0, \quad i = 1, \ldots, q, \quad (6.54)$$
$$g_i(x) \geq 0, \quad i = q + 1, \ldots, m, \quad (6.55)$$

where the following properties hold. (a) The constraints (6.54) include all the nonlinear functions of (C). (b) There is a point x^0 such that $g_i(x^0) > 0$, $i = 1, \ldots, q$, and $g_i(x^0) \geq 0$, $i = q + 1, \ldots, m$. By our Theorem 21 hypothesis such a point exists if there is at least one nonlinear constraint. If there are no nonlinear constraints, and if there are no feasible points interior to some of the linear constraints (that is, $q = 0$), the remarks immediately following can be ignored. (c) For every x^0 where (b) holds,

$$g_i(x^0) \leq 0, \quad i = q + 1, \ldots, m. \quad (6.56)$$

The division of the constraints in this manner is a result of our hypothesis and the concavity of the g_i's. From (b) and (c) it follows that the problem constraints could be written

$$g_i(x) \geq 0, \quad i = 1, \ldots, q, \quad (6.57)$$
$$g_i(x) = 0, \quad i = q + 1, \ldots, m \quad \text{(linear)}. \quad (6.58)$$

6.4 Solution of Convex Problems by Exterior Point Algorithms

We modify the objective function $f(x)$ by adding $\|x - x^*\|^2$ to it, where x^* is the point of Theorem 21 assumed to be an optimum. We then apply our exterior point algorithm to the problem minimize $\tilde{f}(x) = f(x) + \|x - x^*\|^2$ subject to (6.57) and (6.58).

Because the objective function is strictly convex, x^* is the unique solution to the modified programming problem. Those hypotheses of Theorem 27 assumed by Theorem 28 are satisfied if we assume our exterior function to be the quadratic loss function of Section 4.1. Then for any k there is an (x^k, u^k) that is dual feasible [satisfies (6.16) and (6.17)]. We first need to show that every limit point u^* of $\{u^k\}$ is finite.

From Properties b and c there is a point x^0 such that $g_i(x^0) > 0$, $i = 1, \ldots, q$, and $g_i(x^0) = 0$, $i = q + 1, \ldots, m$.

Then, since x^k minimizes the Lagrangian over all x, we have

$$\tilde{f}^k - \sum_{i=1}^{m} u_i^k g_i^k \leq \tilde{f}(x^0) - \sum_{i=1}^{q} u_i^k g_i(x^0) - \sum_{i=q+1}^{m} u_i^k g_i(x^0)$$

$$= \tilde{f}(x^0) - \sum_{i=1}^{q} u_i^k g_i(x^0). \quad (6.59)$$

Taking the limit shows that every u_i^*, a limit point of $\{u_i^k\}$ for $i = 1, \ldots, q$, must be finite; otherwise (6.59) would be violated.

The proof of finiteness for u_i^*, $i = q + 1, \ldots, m$, for which no nonempty interior can be found in common with the constraints $g_1, \ldots, g_q$, is seen as follows.

We need consider only those constraints of (6.55) for which $g_i^k \leq 0$ for k large enough. For those that do not satisfy this, $u_i^k \to 0$. [We point out that $g_i(x^*) = 0$, $i = q + 1, \ldots, m$, but $\lim_{k \to \infty} g_i(x^k)$ may be 0^+.]

For that reduced subset of constraints, because (6.58) is a *consistent* set of linear equality conditions, there is a linearly independent subset of these that generates all the others. We rearrange the reduced set of inequalities from (6.55) so that $g_{q+1}, \ldots, g_{q+p}$ is an independent set of linear constraints, and $g_{q+j}^k \leq 0$, $j = 1, \ldots, p$. Thus for g_{q+p+1}, for example,

$$g_{q+p+1} = \sum_{j=1}^{p} a_{1,j} g_{q+j}.$$

Let A be the unique matrix where $g_{q+p+i} = \sum_{j=1}^{p} a_{i,j} g_{q+j}$, $i = 1, \ldots, r$, where r is the number of remaining dependent constraints of (6.55) for which $g_{q+p+i}^k \leq 0$ for k large.

Because, for the quadratic loss function, $u_i^k = -t_k \, \partial O_i(g_i^k)/\partial g_i = -2t_k(g_i^k - |g_i^k|)/2$, for the dependent constraints under consideration,

$$u_{q+p+i}^k = \sum_{j=1}^{p} a_{i,j} u_{q+j}^k \qquad i = 1, \ldots, r.$$

110   Convex Programming

Substituting this result in the Lagrangian inequality yields

$$\tilde{f}^k - \sum_{i=1}^{m} u_i^k g_i^k \leq \tilde{f}(x) - \sum_{i=1}^{q} u_i^k g_i(x)$$

$$- \left[g_{q+1}(x), \ldots, g_{q+p}(x) \right] \left[I + A^T A \right] \left[u_{q+1}^k, \ldots, u_{q+p}^k \right]^T. \quad (6.60)$$

Since the g_i, $i = q + 1, \ldots, q + p$ are independent linear constraints, and since $[I + A^T A]$ has an inverse, if any $u_i^k \to +\infty$ ($i = q + 1, \ldots, q + p$), x can be selected so that the bound imposed by (6.60) is violated. Thus all multipliers u_i^* are finite.

The assumption that x^* is unique is required so that $\lim_{k \to \infty} x^k = x^*$. Since x^k minimizes $\tilde{\mathcal{L}}(x, u^k)$ everywhere, it follows that $\tilde{f}(x^*) - \sum u_i^* g_i(x^*) = \inf_x [\tilde{f}(x) - \sum u_i^* g_i(x)]$ for every limit point u^* of $\{u^k\}$ and any x.

Making the correspondence $\varphi(x) = f(x) - \sum_{i=1}^{m} u_i^* g_i(x)$, and using Lemma 15, it follows that x^* is an unconstrained minimum of $\mathcal{L}(x, u^*) = f(x) - \sum_{i=1}^{m} u_i^* g_i(x)$. Hence (6.14) is satisfied, as well as (6.10), (6.11), and (6.12), by (x^*, u^*). This proves Theorem 21.   Q.E.D.

We note that the interior of the feasible region R^0 can be empty and exterior point algorithms will still apply.

If $R^0 = \emptyset$, however, the existence of a finite limit for the u_i^k's is not guaranteed. If R^0 is nonempty, then any limit u^* of $\{u^k\}$ must be finite. This can be proved in a manner similar to the proof of finiteness in Theorem 26 for the interior point case, and the proof of Theorem 21 at the end of this section.

6.5 ADDITIONAL RESULTS FOR THE CONVEX ANALYTIC PROBLEM

If the assumptions of Theorem 25 are satisfied, then we have shown that $U(x, r)$ is convex in R^0 for all $r > 0$ and $U(x, r)$ has a finite unconstrained minimum $x(r)$ in R^0 for any $r > 0$. If we further assume that $f(x)$ and $g_i(x)$, $i = 1, \ldots, m$, are analytic in R^0, then by selecting $I_i[g_i(x)]$ to be an analytic function of $g_i(x)$, for all i, it follows that $U(x, r)$ is both convex and analytic in R^0, for every $r > 0$. An interesting consequence of all these assumptions is that $U(x, r)$ must actually be *strictly convex* in R^0 for all $r > 0$. [We do not pursue this analogous development for $T(x, t)$.]

Theorem 30 [Strict Convexity of Analytic Penalty Function]. If in addition to the assumptions of Theorem 25 $U(x, r_k)$ is *analytic* in R^0, then $U(x, r_k)$ is *strictly* convex in R^0 for all $r_k > 0$ and hence has a unique unconstrained minimum for any $r_k > 0$.

6.5 Additional Results for the Convex Analytic Problem

Proof. If $U(x, r_k)$ is *not* strictly convex in R^0, then there is a nondegenerate straight line segment contained in R^0 along which $U(x, r_k)$ varies linearly. Let x^0 and x^1 denote two distinct points on that segment. We assume without loss of generality that $U(x^1, r_k) \leq U(x^0, r_k)$.

We define points on the continuation of the line segment from x^0 through x^1 along which $U(x, r_k)$ varies linearly as

$$x(\lambda) \equiv x^1 + \lambda(x^1 - x^0), \quad \lambda \geq 0.$$

Using the easily proved assertion that the analyticity of $U(x, r_k)$ in R^0 implies that $U(x, r_k)$ remains a linear function along the line segment joining $x(\lambda)$ and x^0 as long as $x(\lambda)$ remains in R^0 yields the following result:

$$U[x(\lambda), r_k] = U(x^1, r_k) + \lambda[U(x^1, r_k) - U(x^0, r_k)] \leq U(x^1, r_k). \quad (6.61)$$

Let $\bar{\lambda}$ be the smallest value for which $x(\lambda) \in R^0$, $0 \leq \lambda < \bar{\lambda}$. By Lemma 12, $\bar{\lambda}$ is finite. Thus $x(\bar{\lambda})$ is on the boundary of R. But because of our assumptions on U this means that

$$\lim_{\lambda \to \bar{\lambda}-} U[x(\lambda), r_k] = +\infty,$$

a contradiction of (6.61).

The uniqueness of the minimum to $U(x, r_k)$ is a direct consequence of the strict convexity. Q.E.D.

Direct Results for Linear Problem

This theorem is applicable to the linear programming problem as a special case. However, for this problem we obtain a stronger and more precise characterization if we take full advantage of the linearity of the problem functions and proceed more directly. In the present context, we formulate a proof in terms of our penalty function.

We assume that $U(x, r) = f(x) + r \sum_{i=1}^{m} I_i[g_i(x)]$, and that $I_i[g_i(x)]$ is differentiable in g_i for $g_i > 0$.

The assumption that the set of points solving the convex programming problem is compact continues to apply. Defining $R_M \equiv \{x \mid f(x) \leq M, x \in R\}$, note that this means that R_M is bounded for every finite M.

We can now prove a result for the linear problem under our usual assumptions, utilizing $U(x, r)$.

Theorem 31 [Existence of Basis for Linear Problem]. If

$$R_M^0 \equiv \{x \mid f(x) < M, g_i(x) > 0; i = 1, \ldots, m\} \neq \emptyset$$

and is bounded for some M, and $f(x), g_1(x), \ldots, g_m(x)$ are linear, then $\Lambda \equiv \{\nabla g_1, \ldots, \nabla g_m\}$ contains n linearly independent vectors.

Proof. The assumptions satisfy the conditions of Theorem 25, so that there exists $x(r) \in R^0$, where $\min_R U(x, r) = U[x(r), r]$. Since $U(x, r_k)$ is differentiable in R^0, this implies that for any $r > 0$

$$\nabla_x U[x(r), r] \equiv \nabla f[x(r)] + r \sum_{i=1}^{m} \frac{\partial I_i\{g_i[x(r)]\}}{\partial g_i} \nabla g_i[x(r)] = 0, \quad (6.62)$$

which means that ∇f is spanned by Λ.

If there are fewer than n linearly independent ∇g_i, then there exists $z \neq 0$ such that $z^T \nabla g_i = 0$, all i. From (6.62) this means that $z^T \nabla f = 0$ also. By hypothesis $R_M^0 \neq \emptyset$ for some $M = M^*$, and so there exists $x^0 \in R_{M^*}^0$. Consider $x^0 + \alpha z$, with α *any* real number. We have that $f(x^0 + \alpha z) = f(x^0) + \alpha z^T \nabla f = f(x^0) < M^*$, since $z^T \nabla f = 0$ and $x^0 \in R_{M^*}^0$. Similarly, $g_i(x^0 + \alpha z) = g_i(x^0) + \alpha z^T \nabla g_i(x^0) > 0$ for $i = 1, \ldots, m$. These inequalities imply that $x^0 + \alpha z \in R_{M^*}^0$.

But since α is arbitrary and $\|z\| \neq 0$ by assumption, this implies that $R_{M^*}^0$ is not bounded, a contradiction. Q.E.D.

If we now make the additional assumption that $I_i[g_i(x)]$ is twice differentiable in g_i for $g_i > 0$, and that $\partial^2 I_i / \partial g_i^2 > 0$ for $g_i > 0$, then we have the following corollary asserting the strict convexity of $U(x, r)$ in R^0 for every $r > 0$ as a direct consequence of Theorem 31 (recall that this result has already been proved as a special case of Theorem 30).

Corollary 21. *If in addition to the assumptions of Theorem 31*

$$\partial^2 I_i(g_i^0) / \partial g_i^2 > 0 \quad (i = 1, \ldots, m) \text{ for } g_i^0 > 0$$

then $U(x, r)$ is strictly convex in R^0 for every $r > 0$.

Proof. Pre- and postmultiplying the matrix of second partial derivatives of $U(x^0, r)$ (where $x^0 \in R^0$, and $r > 0$) by any y yields

$$\sum_{i=1}^{m} \frac{\partial^2 I_i(g_i^0)}{\partial g_i^2} (y^T \nabla g_i)^2. \quad (6.63)$$

Under our assumptions (6.63) is zero only if $y^T \nabla g_i = 0$, $i = 1, \ldots, m$. Since n of the $\{\nabla g_1, \ldots, \nabla g_m\}$ are linearly independent, this means that $y \equiv 0$. Q.E.D.

Note that we have actually obtained the stronger result, that the Hessian of $U(x, r_k)$ is positive definite in R^0 for every $r_k > 0$, under the assumptions of the above theorem and corollary. In particular, as previously observed this implies that the minimizing point $x(r_k)$ is unique in R^0 for each r_k, and $x(r)$ regarded as a function of r is, in fact, an analytic function of $r \in (0, \infty)$ provided that we select I_i to be an analytic function of g_i, for all i.

7

Other Unconstrained Minimization Techniques

In the last four chapters two basic types of unconstrained minimization algorithms were described in detail. Specific examples of interior and exterior point algorithms—the logarithmic penalty function, the inverse penalty function, and the quadratic loss function—have been cited. Obviously many more examples can be generated. In this chapter we describe variations of the basic methods so far discussed and mention other kinds of unconstrained techniques that have computational advantages or intrinsic theoretical interest.

In Section 7.1 it is shown that if a set of values $\{r_1{}^i\}$, $\{t_1{}^i\}$, all i is chosen instead of a single value r_1 to weight the penalty terms attached to the constraints, the initial minimization of the penalty function usually can be facilitated, in a sense to be indicated. This is accomplished by reducing the number of terms in the unconstrained function, as well as by appropriately selecting the weights. The use of quadratic programming algorithms for this purpose is described.

Section 7.2 contains a description of a class of interior point unconstrained algorithms that require no parameter selection. This class of algorithms adds a constraint at each iteration, which requires that the objective function value never exceed that of the current point. An unconstrained function is minimized in the interior of this reduced constraint region. The objective function is then constrained to yield values not exceeding that obtained at the minimizing point, and the unconstrained minimization is applied again. A sequence of strictly decreasing objective function values is computed, and it follows that the minimizing points, under the usual conditions, converge to a local solution of the programming problem B.

For use on analog computers a continuous version of an interior point algorithm is readily derived. An example of this and a description of its

114 Other Unconstrained Minimization Techniques

difference from the Arrow-Hurwicz algorithm are contained in Section 7.3.

Section 7.4 discusses a pure dual algorithm for solving a convex problem with strictly convex objective function. Dependence on the primal variable in the dual problem (of Section 6.2) is eliminated, and the unconstrained functions to be minimized are functions only of the dual variables. It is shown how a perturbation of the objective function makes it possible to solve general convex programming problems with this dual method.

Contrasted with the dual method of Section 7.4, which imposes convexity requirements on the problem functions, is the method of generalized Lagrange multipliers described in Section 7.5, intended to apply to general nonlinear programming problems. The variables may be subject to very general restrictions. In theory, for example, integer programming problems can be solved by this method. The motivation for this algorithm is the interpretation of Lagrange multipliers in resource allocation problems.

In Section 7.6 we discuss an interesting exterior point algorithm, which, for large values of the involved parameter, yields an unconstrained minimum that is a constrained minimum of the original convex programming problem.

Section 7.7 contains a discussion of the theory behind a combined simplicial-unconstrained algorithm, where only the nonlinear constraints are absorbed in the penalty function. An example of this utilizing Rosen's gradient projection method (GP) is discussed in Section 7.8.

7.1 USING WEIGHTED PENALTY FUNCTIONS

Ideally an unconstrained minimization algorithm might have several desirable properties. First, it could take advantage of the initial choice of starting point x^0, in the sense of minimizing the effort to perform the initial minimization of the penalty function. Thus, if a proper choice of parameter values implies that x^0 itself is an unconstrained minimum of the resulting unconstrained function, the algorithm might proceed initially by computing these. Also, if the starting point x^0 is "near" a constrained *solution*, then the algorithm might select the parameters such that the minimizing point of the penalty function would likewise be nearby. The algorithm ideally would anticipate which constraints would not be binding at a solution, and ignore these in a computational procedure. If a solution to an apparently constrained problem is actually unconstrained, it would be very desirable that the algorithm yield a solution of the problem with a single unconstrained minimization of the penalty function.

To these ends, we modify the mixed unconstrained minimization algorithm as follows. The penalty terms I and O are assumed to be separable. We select non-negative values $\{r_k{}^i\}$, $i = 1, \ldots, m$, and $\{t_k{}^i\}$, $i = m + 1, \ldots, q$,

7.1 Using Weighted Penalty Functions

and weight the respective separate penalty terms accordingly, so that the penalty function has the form

$$V(x, \{r_k^i\}, \{t_k^i\}) = f(x) + \sum_{i=1}^{m} r_k^i I_i[g_i(x)] + \sum_{i=m+1}^{q} t_k^i O_i[g_i(x)]. \quad (7.1)$$

This is the mixed interior–exterior point unconstrained minimization function, which applies to Problem M as defined previously (Section 4.3).

It is possible to select the parameters to satisfy the properties indicated above. A revised algorithm proceeds as follows.

Using some selection criterion, select values r_1^i, $i = 1, \ldots, m$, where $r_1^i \geq 0$ (for all i), and values t_1^i, $i = m + 1, \ldots, q$, where $t_1^i \geq 0$ (for all i).

Minimize $V(x, \{r_1^i\}, \{t_1^i\})$ in R^0. If during the course of minimization some g_i becomes infeasible and the associated $r_1^i = 0$, set r_1^i to a strictly positive value and continue to minimize the revised V function, starting from the "last feasible point" in R^0.

For those g_i ($i = m + 1, \ldots, q$), where $g_i(x^0) \geq 0$, clearly, since each $O_i(g_i) = 0$, no t_1^i need be computed, because the respective terms do not contribute to the penalty function. Positive t_1^j are assigned to all those which are infeasible. When any of the currently feasible constraints become infeasible, positive t_1^i are assigned to them according to a specified criterion.

After a number of moves of this type, assume that $V(x, \{r_1^i\}, \{t_1^i\})$ is minimized. At this time all parameters are changed. For all g_i where $1 \leq i \leq m$ and $r_1^i > 0$, the weighting factor r_2^i on $I_i(g_i)$ is selected such that $0 < r_2^i < r_1^i$. For all others, $r_2^i = r_1^i = 0$. For all $g_i \geq 0$, where $m + 1 \leq i \leq q$, let $t_2^i = t_1^i = 0$. For all others in the latter set, the new corresponding weighting factor is increased.

The same general rules apply to the minimization of $V(x, \{r_2^i\}, \{t_2^i\})$. The starting point for the minimization is the unconstrained minimum for the first V function.

This process is continued for sequences of $\{r_k^i\}$, $\{t_k^i\}$, $k = 1, 2, 3, \ldots$, where $\lim_{k \to \infty} r_k^i = 0$, $1 \leq i \leq m$, and $\lim_{k \to \infty} t_k^i = +\infty$, $m + 1 \leq i \leq q$ and $g_i(x^k) < 0$, where x^k denotes an unconstrained minimum in R^0 of

$$V(x, \{r_k^i\}, \{t_k^i\}).$$

This modified algorithm converges to local solutions of Problem M under the same conditions as the mixed algorithm presented in Section 4.3. This occurs because, as before, strict feasibility is still enforced with respect to R^0, and feasibility is eventually enforced with respect to Q.

An interesting question is whether or not criteria for selecting the weighting values can be utilized to reduce the computations required to solve the programming problem. We offer two suggestions of how this can be done and show examples of the efficiency realized.

Criterion 1

Let x^0 be a given point. Initially set $r_1{}^i = 0$, $i = 1, \ldots, m$, and $t_1{}^i = 0$ for $m + 1 \leq i \leq q$ and where $g_i(x^0) \geq 0$. For those other i select $t_1{}^i$ in accordance with the computing rule (to be described). If for some i the $t_1{}^i$ given by the computing rule is 0, set $t_1{}^i =$ some positive value.

Proceeding in accordance with the algorithm, when any g_i ($1 \leq i \leq m$) becomes infeasible, a positive $r_1{}^i$ is attached to the corresponding $I_i(g_i)$ in accordance with the computing rule, and the algorithm proceeds from the "last feasible interior point." If the computing rule yields a value of any $r_1{}^i = 0$, then this parameter is set to a "small" value.

For any i ($m + 1 \leq i \leq q$), if some g_i becomes infeasible, the corresponding $t_1{}^i$ is chosen in accordance with Criterion 2 (or set equal to a small positive value if $t_1{}^i = 0$ as calculated by the computing rule).

Criterion 2

Initially set $t_1{}^i = 0$ ($m + 1 \leq i \leq q$) for all i, where $g_i(x^0) \geq 0$, and choose the remainder in accordance with the computing rule (including the $r_1{}^i$'s).

Later choices for the parameters are made in accordance with Criterion 1.

Computing Rule. *If at some point y a subset of values of $\{r_k{}^i\}$ and $\{t_k{}^i\}$ must be chosen, choose those non-negative values that minimize the norm of the gradient of the V function at that point.*

The computing rule says, in effect: solve a quadratic programming problem of the form

$$\text{minimize } \|b - As\|^2 \tag{7.2}$$

subject to

$$s \geq 0.$$

Choosing the $\{r_k{}^i\}$ and $\{t_k{}^i\}$ in accordance with the computing rule requires solving the above convex programming problem, where the objective function is a positive semidefinite quadratic form and the constraints are simple non-negativity requirements. There are many algorithms that can solve this problem [8, 110, or 78]. The quadratic form in this problem has a special structure. Expanding (7.2), it is clear that the quadratic term is of the form $s^T(A^TA)s$, and the linear term is $-2b^TAs$. Some of these algorithms can be modified to take account of this special structure.

If at some iteration only one weight is required to be chosen, this can be done analytically. The following examples illustrate the revised algorithm.

Example.

$$\text{minimize } x_1 + x_2$$

subject to

$$g_1 = -x_1^2 + x_2 + 2 \geq 0,$$
$$g_2 = -x_1^2 - x_2 + 2 \geq 0,$$
$$g_3 = -x_1 - x_2 + 3 \geq 0.$$

The problem is plotted in Figure 8. We assume a starting point at (0, 0), and the logarithmic penalty function applied to all three constraint functions above.

By Criterion 1, $r_1^i = 0$, $i = 1, 2, 3$. In attempting to minimize $f(x)$ instantaneously, suppose that we choose to make a step down its gradient (1, 1), By whatever process we use to govern step size, eventually we arrive at a point that violates g_1. Backing up, we select some feasible interior point

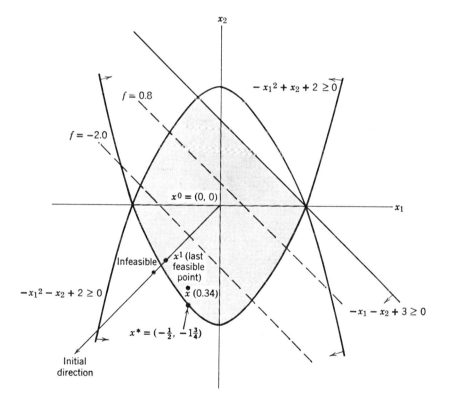

Figure 8 Plot of problem functions for example. The shaded area is the feasible region.

118 Other Unconstrained Minimization Techniques

along that ray, say at $(-1 + \epsilon, -1 + \epsilon)$, where $\epsilon > 0$. Let $\epsilon = 0.1$, giving $g_1 = 0.29$, $\nabla g_1 = (+1.8, +1)$.

The computing rule requires minimization of the norm of the gradient of the penalty function. For this example the gradient at $(-0.9, -0.9)$ is

$$\begin{pmatrix} 1 \\ 1 \end{pmatrix} - \left(\frac{r_1^1}{(0.29)}\right) \begin{pmatrix} 1.8 \\ 1.0 \end{pmatrix}.$$

Now we solve the quadratic programming problem

$$\text{minimize} \left\| \begin{pmatrix} 1 \\ 1 \end{pmatrix} - r_1^1 \begin{pmatrix} 6.21 \\ 3.45 \end{pmatrix} \right\|^2$$

subject to

$$r_1^1 \geq 0.$$

The r_1^1 that solves this problem is 0.34. Our unconstrained function is now

$$x_1 + x_2 - (0.34) \ln(-x_1^2 + x_2 + 2)$$

and we are at $(-0.9, -0.9)$. Using the usual analytic means, we compute the minimizing point,

$$x_1(0.34) = -0.50, \quad x_2(0.34) = -1.41.$$

Since $(-0.50, -1.41)$ is feasible with respect to g_2 and g_3, if V had been minimized by numeric means (see Chapter 8) r_1^2 and r_1^3 would never have been introduced. Thus the computational effort is considerably reduced from our original method, which would have assigned positive values and minimized a more difficult function. In this example nonbinding constraints were identified properly for the first value of r.

Choosing the last feasible interior point along a ray that penetrates a constraint boundary is not as arbitrary as it might seem.

Let $V(x, \{r_1^i\}, \{t_1^i\}) = V_1(x)$ denote the current form of the mixed penalty function. Let a denote the index of the constraint to be added to V_1, let z denote the point where the ray emanating from some point x^0 interior to g_a pierces its boundary, and let $\tilde{x}$ denote the last interior feasible point. Then $\tilde{x} = x^0 + \lambda(z - x^0)$, where λ is presumed close to 1.

The problem of selecting r_1^a according to the computing rule is

$$\text{minimize} \left\| \nabla V_1(\tilde{x}) - \left(\frac{r_1^a}{g_a(\tilde{x})}\right) \nabla g_a(\tilde{x}) \right\|^2$$

subject to

$$r_1^a \geq 0.$$

7.1 Using Weighted Penalty Functions

Obviously, for $\tilde{x}$ close to z, since $\nabla V_1(z)$ is assumed finite, the optimal r_1^a, which is given by

$$r_1^a(\tilde{x}) = \frac{\nabla^T V_1(\tilde{x}) \nabla g_a(\tilde{x})}{\|\nabla g_a(\tilde{x})\|^2} g_a(\tilde{x}), \qquad (7.3)$$

must be positive. This is because, if the inner product in the numerator of (7.3) were negative, the current unconstrained function would generally be minimized at a point interior to g_a and it would not have to be added to the penalty function. It is clear from (7.3) that the ratio $r_1^a(\tilde{x})/g_a(\tilde{x})$ approaches a constant as $\tilde{x} \to z (\lambda \to 1)$. That constant is

$$w_a = \frac{\nabla V_1^T(z) \nabla g_a(z)}{\|\nabla g_a(z)\|^2}. \qquad (7.4)$$

[Recall, since g_a is concave, that if $\nabla g_a(x) = 0$ for any x, at a point where $g(x) = 0$, no interior exists to the region $g_a(x) \geq 0$. Hence we can assume $\nabla g_a(z) \neq 0$, and thus w_a is well defined.]

For the logarithmic penalty function, as noted at the end of Section 6, it is possible to preset the difference between the primal and dual objective function values at the first unconstrained minima by making the appropriate r_1 selection. Obviously the same remark applies to the selection of a different r_1^i, one for each constraint. Thus the choice of how close to the boundary of g_a the last interior point is to be selected can be restated in terms of how close the user wishes to estimate the optimum solution value at the first unconstrained minimum. Suppose that the tolerance is $\epsilon > 0$. Then w_a should be approximated [any point near z will give a good approximation to (7.4)] and λ chosen so that

$$g_a(\tilde{x}) = g[x^0 + \lambda(z - x^0)] = \frac{\epsilon/m}{w_a}.$$

Then r_1^a is given as

$$r_1^a = g_a(\tilde{x}) w_a.$$

Then the contribution of the ath penalty term to the difference between the primal and dual values will be

$$\frac{r_1^a}{g_a(x^1)} g_a(x^1) = g_a(\tilde{x}) w_a \doteq \frac{\epsilon}{m},$$

where x^1 is the unconstrained minimum of $V(x, \{r_1^i\}, \{t_1^i\})$.

Using selection criterion on another problem, we see how initial choosing of the proper r values leads to computational efficiencies.

120 Other Unconstrained Minimization Techniques

Example. The problem is

$$\text{minimize } x_2$$

subject to

$$g_1 = -(x_1 + 1)^2 - x_2^2 + 4 \geq 0,$$
$$g_2 = -(x_1 - 1)^2 - x_2^2 + 4 \geq 0,$$
$$g_3 = -x_1 - x_2 + 2 \geq 0.$$

The interior starting point is given as $(x_1^0, x_2^0) = (0, -1)$.

We note that, regardless of the particular penalty functions used, the associated quadratic programming algorithm is independent of it in the following sense. That problem is

$$\text{minimize } \|\nabla f^0 - N^0 w\|^2$$

subject to

$$w \geq 0,$$

where N^0 is the matrix of relevant constraint gradients evaluated at x^0. If w^0 is the solution of that problem, then

$$r_1^i = \frac{-w_i^0}{\partial I_i^0 / \partial g_i} \quad \text{and} \quad t_1^i = \frac{w_i^0}{\partial O_i^0 / \partial g_i}.$$

We assume that an interior point algorithm applies to all three constraints. Then for the previous example we must solve the quadratic problem

$$\text{minimize } \left\| \begin{pmatrix} 0 \\ 1 \end{pmatrix} - \begin{Bmatrix} -2 & 2 & -1 \\ 2 & 2 & -1 \end{Bmatrix} \begin{pmatrix} w_1 \\ w_2 \\ w_3 \end{pmatrix} \right\|^2$$

subject to

$$w_1, w_2, w_3 \geq 0.$$

The optimum choice of w_i^0's is $(\frac{1}{4}, \frac{1}{4}, 0)$. If we use the logarithmic penalty function for all three functions, then, since $\partial I_i / \partial g_i = -1/g_i$,

$$r_1^1 = [g_1(x^0)](\tfrac{1}{4}) = \tfrac{1}{2},$$
$$r_1^2 = [g_2(x^0)](\tfrac{1}{4}) = \tfrac{1}{2},$$
$$r_1^3 = [g_3(x^0)](0) = 0.$$

If we use the P function,

$$r_1^1 = (g_1^0)^2(\tfrac{1}{4}) = 1,$$
$$r_2^1 = (g_2^0)^2(\tfrac{1}{4}) = 1,$$
$$r_3^1 = (g_3^0)^2(0) = 0.$$

7.2 Q-Function Interior Point Minimization Algorithms

Several interesting results follow regardless of which of these two penalty terms we choose.

The third constraint, which is not binding at the solution, has been given a zero weight and thus will never enter into the calculations.

In this example these choices of r_1^i yield a value of zero for the norm of the gradient of the unconstrained function. Since this is a convex programming problem, we are at an unconstrained minimum to start with, and thus have a *dual feasible point*.

Both the logarithmic and P-penalty functions give a dual value of $f - \sum w_i g_i$. Thus if v^* is the optimum value we have already bracketed it, since

$$f - \sum w_i g_i = -1 - (\tfrac{1}{4})(2) - (\tfrac{1}{4})(2) - (0)(1) = -2 \leq v^* \leq -1 = f.$$

The solution value of this problem is ~ -1.732, which clearly lies in that range.

In this problem the selection of initial r values using quadratic programming is of twofold benefit. These advantages are much more significant for larger problems.

7.2 Q-FUNCTION TYPE INTERIOR POINT UNCONSTRAINED MINIMIZATION ALGORITHMS

For programming problems of type (D)— minimize $f(x)$ subject to $g_i(x) \geq 0$, $i = 1, \ldots, m$, where $R^0 = \{x \mid g_i(x) > 0, i = 1, \ldots, m\}$ is a nonempty set—it is possible to describe a general class of interior point algorithms that have the following properties. (a) The selection of the strictly decreasing sequence $\{r_k\}$ of non-negative parameters used to weight the usual interior penalty function is not required. (b) Each unconstrained minimization (say at iteration k) depends only on the value of the objective function at the starting point, x^{k-1}, an unconstrained minimum for iteration $k - 1$. (c) The sequence of interior feasible points has objective function values $f^0, f^1, f^2, \ldots, f^k$, which form a strictly decreasing sequence, starting from the initial point. Under the usual assumptions these values converge locally. A general class of procedures closely related to those presented in this section, based essentially on treating the objective function as a moving constraint, has been dubbed the "method of centers" by Huard [73] and has been further developed by Bui Trong Lieu and Huard [16] for general spaces. (This was discussed briefly in the historical survey given in Chapter 1.)

Other advantages are retained for the convex programming problem—global convergence—and generation of dual feasible points, which, in the limit as $k \to \infty$, solve the (Wolfe) dual convex programming problem.

Crucial to the development is the characterization of a function, hereafter called a Q function, defined in the following way.

Let $z = (z_0, z_1, \ldots, z_m)^T$ be an $m + 1$ dimensional positive vector. The function $Q(z)$ is a Q function when (a) $Q(z)$ is continuous when $z > 0$, and (b) if $\{z^i\}$ is an infinite sequence of vectors, where $z^i > 0$ for all i and $\bar{z}$ is the limit of $\{z^i\}$ with the property that $\bar{z}_j = 0$ for some j, then $\liminf_{i \to \infty} Q(z^i) = \alpha$ (possibly infinite), where $\alpha > Q(z)$ for all $z > 0$.

The steps of the algorithm are as follows.

1. Let $x \in R^0$ be the starting point for the process.
2. Let $R_1 = \{x \mid f(x) \leq f(x^0); x \subset R\}$.
3. Define $Q^1(x) = Q[-f(x) + f(x^0), g_1(x), \ldots, g_m(x)]$, where Q is a Q function.
4. Find a local minimum of $Q^1(x)$ in R_1. Any existing local minimum is unconstrained, since all boundary points of R_1 give higher values of $Q^1(x)$ than interior points of R_1.
5. Let $R_k = \{x \mid f(x) \leq f(x^{k-1}); x \in R\}$. Minimize

$$Q^k(x) = Q[-f(x) + f(x^{k-1}), g_1(x), \ldots, g_m(x)]$$

in R_k for $k = 1, 2, 3, \ldots$, thus generating a minimizing sequence $\{x^k\}$.

6. Clearly, if $R_k^0 \neq \emptyset$, then $f(x^k) < f(x^{k-1})$. The process should approach some local minimum of f in R.

An example of a Q function is provided by

$$Q^k(x) = \frac{1}{f(x^k) - f(x)} + \sum_{i=1}^{m} \frac{1}{g_i(x)}.$$

This function was developed in detail by the authors [52] to provide a method of the class described above for the convex programming problem.

It may be noted that the above Q function is closely related to the interior point function $P(x, r)$ (3.10). In fact, we have shown [52] that the minimizing sequence of Q corresponds to the minimizing sequence of P for a specified sequence of values of the parameter r.

The conditions under which the algorithm converges are similar to those for general interior point unconstrained minimization algorithms given in Section 3.3, Theorem 8.

Theorem 32 [Convergence of Q-Function Algorithms to Compact Sets of Local Solutions]. If (a) the functions $f, g_1, \ldots, g_m$ are continuous, (b) R^0 is nonempty, (c) $A^* \cap R^0 \neq \phi$ (where A^* is as follows), (d) a set of points A^*

7.2 Q-Function Interior Point Minimization Algorithms

that are local minima of Problem B with local minimum value v^* is a nonempty isolated compact set, and (e) the function Q is a Q function, then

(i) if $R_k^0 \neq \emptyset$, there is a point x^k that is a local unconstrained minimum of Q^k in R_k^0, and every limit point of the uniformly bounded sequence $\{x^k\}$ is a local minimum with local minimum value v^*, and

(ii) if $R_k^0 = \emptyset$ for some finite k, x^{k-1} is a local minimum with value v^*, that is, x^{k-1} is an unconstrained solution of (B).

Proof. Let x^0 be a starting point in the interior of the set $R \cap S$, where S is the compact perturbation set given by Theorem 7. Unless $f(x^0) = v^*$, $R_1^0 \cap S$ is nonempty. By a trivial modification of Corollary 8 and the definition of $Q^1(x)$, there is an unconstrained minimum x^1 in $R_1^0 \cap S$. Since $R_1^0 \cap S \subset R_1^0$, x^1 is a *local* unconstrained minimum of Q^1 in R_1^0. The same remarks hold for all k for which $R_k^0 \neq \emptyset$.

If, for some k, $R_k^0 = \emptyset$, this means that there is no x such that $x \in R^0 \cap S$ and $f(x) < f(x^{k-1})$, with $x^{k-1} \in R^0$. Then x^{k-1} must be a local constrained minimum of f in R. Since $x^{k-1} \in R^0$, this point is an unconstrained minimum of $f(x)$, and since $x^{k-1} \in S$, from Theorem 7 we have $f(x^{k-1}) = v^*$.

The compactness of S implies that $\{x^k\}$ has a convergent subsequence, which we still denote by $\{x^k\}$. We can assume, therefore, that $x^k \to \bar{x} \in R \cap S$.

If $R_k^0 \neq \emptyset$ for all k, then $f(x^k) < f(x^{k-1})$. Thus the sequence of values $\{f^k\}$ is a strictly decreasing sequence and has a limiting value $\bar{v} \geq v^*$. We must show that $\bar{v} = v^*$.

Assume $\bar{v} > v^*$. Since the value of Q^k at its minimum is $Q^k(x^k) = Q[f(x^k) + f(x^{k-1}), g_1(x^k), \ldots, g_m(x^k)]$, and $[-f(x^k) + f(x^{k-1})] \to 0$ as $k \to \infty$ and $g_i(x^k) \to g_i(\bar{x})$, all i, from the definition of the Q function it follows that $\liminf_{k \to \infty} Q^k(x^k) = \alpha$, where $\alpha > Q^k(x^0)$ for any $x^0 \in R^0 \cap S$. Let x^0 be such that $x^0 \in R^0 \cap S$, and $v^* < f(x^0) < \bar{v}$. Such an x^0 exists by our present assumption and Hypothesis c. Note that $x^0 \in R_k^0$ for all k. Thus, if $\bar{v} > v^*$, there exists a point in $R_k^0 \cap S$ giving a Q^k value less than $Q^k(x^k)$ when k is large. This contradicts the fact that x^k minimizes $Q^k(x)$ in $R_k^0 \cap S$.

Q.E.D

We give an example of this algorithm.

Example.

$$\text{minimize } x_1 x_2$$

subject to

$$x_1 \geq 0,$$
$$x_2 \geq 0.$$

Suppose $x^0 = (\tfrac{1}{2}, 2)$ is the given interior starting point. We add the constraint

$$f(x^0) - x_1 x_2 = 1 - x_1 x_2 \geq 0$$

and apply the logarithmic penalty function to all three constraints. This yields

$$Q^1 = -\ln(1 - x_1 x_2) - \ln(x_1) - \ln(x_2)$$

as the Q function to be minimized. Differentiating and solving yields $x^1 = (1/\sqrt{2}, 1/\sqrt{2})$. We now add the constraint

$$\tfrac{1}{2} - x_1 x_2 \geq 0,$$

which yields

$$Q^2(x) = -\ln(\tfrac{1}{2} - x_1 x_2) - \ln(x_1) - \ln(x_2).$$

The solution is $x^2 = (\tfrac{1}{2}, \tfrac{1}{2})$. Clearly $x^k = (2^{-k/2}, 2^{-k/2})$, which approaches $(0, 0)$ as $k \to \infty$. This example is plotted in Figure 9.

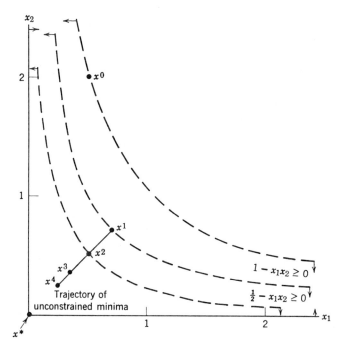

Figure 9 Example of nonlinear programming solution by Q-function type unconstrained algorithms. The problem is:

$$\text{minimize } x_1 x_2$$

subject to

$$x_1 \geq 0$$
$$x_2 \geq 0$$
$$Q^{k+1} = -\ln(f^k - x_1 x_2) - \ln x_1 - \ln x_2.$$

7.2 Q-Function Interior Point Minimization Algorithms

For convex programming problems, each constraint $-f(x) + f(x^k)$ is a concave function constrained to be non-negative, and hence defines a convex region. Thus if the Q function is selected properly, that is, if it is a convex decreasing function of each of its arguments, it is convex in x. This development can be pursued in a manner identical with that in Section 6.3. Thus we can select each Q^k as a convex function in R_k^0, implying that local unconstrained minima of Q^k are global unconstrained minima. Every x^k yields a dual feasible point. This is seen as follows.

Suppose that each Q^k is separable in its arguments and is given by

$$Q^k = Q_k^0(-f + f^{k-1}) + \sum_{i=1}^m I_i[g_i(x)],$$

where $Q_k^0(-f + f^{k-1})$, $I_i(g_i)$ ($i = 1, \ldots, m$) are convex, decreasing, and differentiable in their arguments. Obviously, then, each Q^k is convex. Let

$$u_i^k = \frac{\partial I_i[g_i(x^k)]/\partial g_i}{\partial Q_k^0[-f(x^k) + f(x^{k-1})]/\partial f}, \qquad i = 1, \ldots, m.$$

Then (x^k, u^k) is feasible for the dual convex programming problem [(6.16) and (6.17)] and, as $k \to \infty$, yields limit points that solve that problem.

The proof of this assertion is essentially that given in Section 6.3 for general interior point unconstrained minimization algorithms and will not be repeated here. If the functions $f, g_1, \ldots, g_m$ are continuously differentiable, the proof, using (6.15) as a dual constraint, is obvious. That is, since x^k is an unconstrained minimum,

$$\nabla Q^k(x^k) = 0 = -\frac{\partial Q_k^0[-f(x^k) + f(x^{k-1})]}{\partial f} \nabla f(x^k) + \sum_{i=1}^m \frac{\partial I_i[g_i(x^k)]}{\partial g_i} \nabla g_i(x^k).$$

Multiplying by $-\{\partial Q_k^0[-f(x^k) + f(x^{k-1})]/\partial f\}^{-1}$ yields an equation of the form of (6.15). The non-negativity of each u_i^k follows from the decreasing property assumed for each of the terms.

Thus Q-function algorithms have, for the convex programming problem, all the advantages of the penalty function interior point algorithms.

We have already indicated a connection between the minimizing sequences generated by a specific method of centers type Q function and an interior point U function. This connection has been derived and established for a large class of Q and U functions by Fiacco [48]. It is shown that such functions can be functionally related in a very simple way, leading to an explicit relation among the respective minimizing points. The resulting sequential procedures are essentially equivalent under certain conditions. It is of further interest that the analogous correspondence is developed for

126   Other Unconstrained Minimization Techniques

exterior point methods in [48], leading to a class of new procedures that may aptly be described as "exterior centers methods."

Further connections analogous to the above have also been indicated in results recently obtained by Falk [42].

7.3 CONTINUOUS VERSION OF INTERIOR POINT TECHNIQUES

A natural variation of interior point unconstrained techniques is to let the parameter r vary continuously as a decreasing function of time and solve a differential equation of the form $dx/dt = -\nabla_x U(x, t)$. (Considered here is Problem B with inequality constraints only.) Motion is analogous to Cauchy's method of steepest descent (Section 8.2). Such a method is suitable for use on an analog computer.

For definiteness let $r = e^{-t}$, and let U be the P function of (3.10). Then $P(x, t) = f(x) + e^{-t} \sum_{i=1}^{m} 1/g_i(x)$, and

$$\dot{x} = -\left[\nabla f(x) - e^{-t} \sum_{i=1}^{m} \frac{1}{g_i^2(x)} \nabla g_i(x)\right]. \tag{7.5}$$

One can, under the same general conditions of Theorem 8, Section 3.3, prove that solving (7.5) results in the generation of points satisfying the first-order conditions of (2.5), (2.7), (2.8), and (2.9), Section 2.1. This is done in Theorem 33. Fortunately, the first-order necessary conditions are sufficient that a point be a global solution of the convex programming problem. Hence in Theorem 34 convergence to a global solution is proved.

Theorem 33 [Stability of Stationary Points of the Lagrangian]. If (a) f, $g_1, \ldots, g_m$ are continuously differentiable functions of x, (b) R^0 is nonempty, (c) the closure of R^0 is R, and (d) a set of points A^* that are local minima corresponding to the local minimum value v^* is a nonempty isolated compact set, then there exists a set $N^* \supset A^*$ and a value $t^0 > 0$ such that, for the starting point (x^0, t^0), where $x^0 \in N^*$, every limiting point x^* of the solution of (7.5) satisfies (2.5), (2.7), (2.8), and (2.9) (where the corresponding u^* is also generated by the differential equation). Note that u^* is not necessarily finite.

Proof. By the chain rule,

$$\frac{dP}{dt} = -|\nabla P[x(t), t)]|^2 - e^{-t} \sum_{i=1}^{m} \frac{1}{g_i[x(t)]} < 0 \tag{7.6}$$

when $x(t) \in R^0$. Thus if $x(0) \in R^0$, the trajectory generated by (7.5) can never

7.3 Continuous Version of Interior Point Techniques

leave the feasible region, since P would have infinite value there, contradicting the strictly monotonic decreasing property ensured by (7.6).

Let S be the compact perturbation set (about A^*) guaranteed by Theorem 7 (because of our Hypothesis d). By continuity and the definition of $P(x, t)$ it can be assumed that $P(x, t) \geq v^* + \lambda$ on the boundary of $S \cap R$ for some $\lambda > 0$. Let $N^* = S \cap R^0 \cap \{x \mid f(x) \leq v^* + \lambda/2\}$. The set N^* is not empty by our Assumptions a, b, and c. (Equation 7.5 is well defined by Assumption a). Let t^0 be such that $e^{-t^0} \sum_{i=1}^m 1/g_i(x^0) < \lambda/2$. Since $P[x(t), t]$ decreases strictly, $x(t)$ can never exit from the compact set $S \cap R$. Expanding P in a Taylor series,

$$P[x(2t), 2t] - P[x(t), t] = t \left\{ -|\nabla P[x(\xi), \xi]|^2 - e^{-\xi} \sum_{i=1}^m \frac{1}{g_i[x(\xi)]} \right\}, \quad (7.7)$$

where $t \leq \xi \leq 2t$.

Since the left-hand side goes to zero as $t \to \infty$, each term in brackets on the right-hand side of (7.7) must go to zero. Let

$$u_i(t) = \frac{e^{-t}}{g_i^2[x(t)]}, \quad i = 1, \ldots, m. \quad (7.8)$$

Then every limit point of $[x(t), u(t)]$ satisfies the first-order necessary conditions of Section 2.1 for a local constrained minimum. Q.E.D.

In general, then, (7.5) can lead only to a stationary point (in x) of the Lagrangian function $\mathcal{L}(x, u) = f(x) - \sum u_i g_i(x)$. For convex programming problems this fact, along with complementarity, is enough to ensure convergence to a global solution.

Theorem 34 [Global Stability for the Convex Programming Problem]. If (a) $f, -g_1, \ldots, g_m$ are convex continuously differentiable function of x, (b) R^0 is nonempty, and (c) the set of points that solve the convex programming problem is bounded (and hence compact), then from any starting point (x^0, t^0), where $x^0 \in R^0$, every limit point of $[x(t), u(t)]$ [where $u(t)$ is given by (7.8)] satisfies the sufficient conditions for a minimum to the convex programming problem (6.10–6.13).

Proof. Let x^0 be any point in R^0, and t^0 be any value (possibly negative). From our Assumptions a–c and Theorem 24, if $x(t)$ has an unbounded limit point and remains feasible, $f[x(t)] \to +\infty$. But, using the analogous steps of the proof of Theorem 33, P strictly decreases and hence $x(t)$ remains in R^0. Thus $x(t)$ must have bounded limit points; otherwise, since $P > f$,

$$P[x(t), t] \to +\infty.$$

By the same arguments as those used in the proof of Theorem 26, every limit point of $u(t)$ is finite. Thus, since

$$\lim_{t \to \infty} \left| \nabla f[x(t)] - \sum_{i=1}^{m} u_i(t) \nabla g_i[x(t)] \right|^2 + \sum u_i(t) g_i[x(t)] = 0,$$

every limit point of $[x(t), u(t)]$ satisfies the sufficiency conditions (6.10–6.13) that a limit point of $x(t)$ be a minimum of the convex programming problem.
Q.E.D.

The following example, solved on an analog computer, illustrates the use of (7.5).

Example.

$$\text{minimize} \sum_{j=1}^{5} e_j x_j + \sum_{i=1}^{5} \sum_{j=1}^{5} c_{ij} x_i x_j + \sum_{j=1}^{5} d_j x_j^3$$

subject to

$$g_i = \sum_{j=1}^{5} a_{ij} x_j + b_i \geq 0, \, i = 1, \ldots, 10, \quad x_j \geq 0, j = 1, \ldots, 5.$$

When $[c_{ij}]$ is a positive semidefinite matrix, and $d_j \geq 0, j = 1, \ldots, 5$, the objective function is convex in the feasible region. Using data given in [85], page 18, this example was solved on an analog computer [13]. The theoretical solution is approximately $x^* = (0.300, 0.333, 0.400, 0.428, 0.224)$.

In Figure 10 the components of the solution of (7.5) are graphed, starting from $x^0 = (0.1, 0.2, 0.2, 0.5, 0.5)$. Equilibrium was reached at 0.8 seconds.

The accuracy of the analog solution is considerably less than that obtained by a digital computer (see [49] for the digital solution). For some problems this may be offset by the flexibility of altering the inputs once the problem setup has been accomplished.

Comparison with Classical Arrow-Hurwicz Results

In a paper in 1951 [6], Arrow and Hurwicz proposed a "differential" method for solving convex programming problems. If the problem is written as Problem C their method is essentially as follows. Find a saddle point in (x, u) of the Lagrangian function $\mathcal{L}(x, u) = f(x) - \sum u_i g_i(x)$. If there is a point $(\bar{x}, \bar{u})$ such that $\mathcal{L}(\bar{x}, u) \leq \mathcal{L}(\bar{x}, \bar{u}) \leq \mathcal{L}(x, \bar{u})$ for all $u \geq 0$, then $\bar{x}$ solves the convex programming problem (C). This is easily seen by relating this to the sufficiency conditions (6.10–6.13).

7.3 Continuous Version of Interior Point Techniques

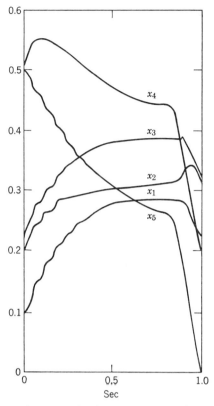

Figure 10 Analog computer solution.

The differential equations of Arrow and Hurwicz are given as

$$\dot{x} = -\nabla_x \mathcal{L}(x, u), \tag{7.9}$$

$$\dot{u}_i = 0 \quad \text{if} \quad u_i = 0, \quad g_i(x) > 0,$$

and

$$\dot{u}_i = -g_i(x) \quad \text{otherwise}, \quad i = 1, \ldots, m. \tag{7.10}$$

Thus there is independent variation of both x and u, as opposed to the method presented in (7.5). In the latter method, as seen in (7.8), any u_i is generated explicitly in the differential equation form of the P function as the ratio $e^{-t}/g_i^2[x(t)]$.

In order to prove convergence of (7.9) and (7.10), Arrow and Hurwicz [5] showed that if the matrix of second partial derivatives of the *Lagrangian* is positive definite, solution of (7.9) and (7.10) generates a minimum of the convex programming problem.

The requirements for convergence of (7.5) are less stringent, as seen in Theorem 34.

7.4 DUAL METHOD FOR STRICTLY CONVEX PROBLEMS

A much more general version of a "dual method" for strictly convex programming problems is contained in [41] and [43]. The method has interest for several reasons. First, it is an unconstrained minimization technique but can be regarded as unconstrained in the dual variables $\{u_i\}$ rather than the primal variables $\{x_j\}$. In effect it solves the dual convex programming problem of Section 6.2. The class of problems to which it applies are called "strictly convex" in that the Lagrangian $\mathcal{L}(x, u)$ is strictly convex in x for the dual variables $\{u_i^*\}$, which are given by Theorem 1. Second, it is a technique that "iterates" rather than employs differential equations. It is therefore amenable to solution on a digital computer. Its relationship to the Arrow-Hurwicz technique is essentially the same as that of the iterative version of the P function to the continuous analog of Section 7.3. Third, at the end of this section it is shown how an ϵ-perturbation technique makes the method applicable to all convex problems.

Rewriting the primal and dual convex programming problems of Section 6.2, we have

PRIMAL

$$\text{minimize } f(x)$$

subject to

$$g_i(x) \geq 0, \quad i = 1, \ldots, m. \tag{7.11}$$

DUAL

$$\text{maximize } \mathcal{L}(x, u) \equiv f(x) - \sum_{i=1}^{m} u_i g_i(x)$$

subject to

$$\mathcal{L}(x, u) \equiv \inf_{y} \mathcal{L}(y, u), \tag{7.12}$$

$$u_i \geq 0, \quad i = 1, \ldots, m. \tag{7.13}$$

Now if for any $u \geq 0$ the $x(u)$ that satisfies (7.12) is unique, it can be substituted into the dual objective function, yielding the following programming problem:

$$\text{maximize } \gamma(u) = \mathcal{L}[x(u), u] = f[x(u)] - \sum_{i=1}^{m} u_i g_i[x(u)] \tag{β}$$

subject to

$$u \in D(\gamma), \tag{7.14}$$

7.4 Dual Method for Strictly Convex Problems

where $D(\gamma) = \{u \mid u \geq 0, \mathcal{L}(x, u) \text{ is strictly convex, and a finite } x \text{ exists such that } (x, u) \text{ satisfies (7.12)}\}$. (Note that the x must therefore be unique.)

Problem β is a more satisfactory dual in the sense that it depends only on the dual variables $\{u_i\}$. Also, under appropriate assumptions, it is a convex programming problem [that is, $\gamma(u)$ is concave and $D(\gamma)$ is a convex set], and the solution value coincides with that of the primal convex programming problem.

Theorem 35 [Solution of Strictly Convex Problem by Dual Lagrange Multiplier Algorithm]. If $f, -g_1, \ldots, -g_m$ are convex functions, and if, associated with x^*, a solution to the primal convex programming problem is a vector u^* satisfying (6.10), (6.11), (6.12), and (6.14) (the sufficiency conditions of Theorem 19) and $u^* \in D(\gamma)$, then

(i) x^* is a *unique* solution of the primal problem,
(ii) $D(\gamma)$ is a *nonempty open set* [relative to $(E^m)^+$],
(iii) $D(\gamma)$ is a convex set,
(iv) $\gamma(u)$ is concave over $D(\gamma)$, and
(v) u^* solves Problem β.

Proof. Since u^* is assumed in $D(\gamma)$, Part i is proved. To prove Part ii we need to show that (a) in a non-negative neighborhood about u^*, $\mathcal{L}(x, u)$ is strictly convex, and (b) $\mathcal{L}(x, u)$ is minimized at a finite point for all u in this neighborhood. To prove (a), let x^1 and x^2 be any two distinct points, and let $\Delta u \equiv (\delta u_i, \ldots, \delta u_m)$ be small enough so that $u_i^* + \delta u_i > 0$ for all i, where $u_i^* > 0$. Let α be a scalar where $0 < \alpha < 1$. Then

$$\mathcal{L}[\alpha x^1 + (1-\alpha)x^2, u^* + \Delta u] = f[\alpha x^1 + (1-\alpha)x^2]$$

$$- \sum_{i=1}^{m}(u_i^* + \delta u_i)g_i[\alpha x^1 + (1-\alpha)x^2].$$

Let $Z = \{i \mid u_i^* = 0\}$. Now if for any i, where $u_i^* > 0$ ($i \notin Z$),

$$g_i[\alpha x^1 + (1-\alpha)x^2] > \alpha g_i(x^1) + (1-\alpha)g(x^2),$$

we are done, since $(u_i^* + \delta u_i) > 0$ for that i. Hence the worst that can happen is that $g_i[\alpha x^1 + (1-\alpha)x^2] = \alpha g_i(x^1) + (1-\alpha)g_i(x^2)$. For those i's where $u_i^* = 0$ ($i \in Z$), $\delta u_i \geq 0$, and that term contributes to the following inequality to our benefit. Taking the worst case and regrouping, we have

$$\mathcal{L}[\alpha x^1 + (1-\alpha)x^2, u^*] - \sum \delta u_i g_i[\alpha x^1 + (1-\alpha)x^2]$$

$$< \alpha \mathcal{L}(x^1, u^*) + (1-\alpha)\mathcal{L}(x^2, u^*) - \sum \delta u_i g_i[\alpha x^1 + (1-\alpha)x^2]$$

[by the strict convexity of $\mathcal{L}(x, u^*)$]

$$= \alpha\mathcal{L}(x^1, u^*) + (1 - \alpha)\mathcal{L}(x^2, u^*) - \sum_{i \notin Z} \delta u_i[\alpha g_i(x^1) + (1 - \alpha)g_i(x^2)]$$
$$- \sum_{i \in Z} \delta u_i\{g_i[\alpha x^1 + (1 - \alpha)x^2]\}$$

(by our assumption about the i's not in Z)

$$\leq \alpha\mathcal{L}(x^1, u^*) + (1 - \alpha)\mathcal{L}(x^2, u^*) - \sum_{i \notin Z} \delta u_i[\alpha g_i(x^1) + (1 - \alpha)g_i(x^2)]$$
$$- \sum_{i \in Z} \delta u_i[\alpha g_i(x^1) + (1 - \alpha)g_i(x^2)], \quad \delta u_i \geq 0 \text{ for } i \in Z,$$
$$= \alpha\mathcal{L}(x^1, u^* + \Delta u) + (1 - \alpha)\mathcal{L}(x^2, u^* + \Delta u).$$

This proves Part (a).

Let v^* denote $\mathcal{L}(x^*, u^*)$. Let $N(x^*, \delta)$ be a neighborhood about x^* of radius $\delta > 0$. Let $v_1 = \inf_B \mathcal{L}(x, u^*)$, where B is the boundary of $N(x^*, \delta)$. Because $\mathcal{L}(x, u^*)$ is strictly convex, $v_1 > v^*$. Clearly, then, we can shrink the set of u's about u^* for which $\mathcal{L}$ is strictly convex so that there is an $\epsilon > 0$ such that for all $u \in \{u = u^* + \Delta u \mid \|\Delta u\| < \epsilon\} \cap E_m^+$,

$$\inf_B \mathcal{L}(x, u^* + \Delta u) > \mathcal{L}(x^*, u^* + \Delta u).$$

[Otherwise there exists a point $x_B \in B$ where $\mathcal{L}(x_B, u^*) \leq \mathcal{L}(x^*, u^*)$, contradicting the strict convexity of $\mathcal{L}(x, u^*)$.] Then $\mathcal{L}(x, u)$ must take on its infimum in the interior of $N(x^*, \delta)$ and not on its boundary B. Thus $D(\gamma)$ is nonempty and, since the same argument applies to any $u \in D(\gamma)$, $D(\gamma)$ is open with respect to $(E^m)^+$.

To prove (iii), let u^1 and u^2 be two vectors in $D(\gamma)$. Let x^1 and x^2 be the two unique vectors where (x^1, u^1) and (x^2, u^2) satisfy (7.12). [Note that

$$\mathcal{L}[x, \alpha u^1 + (1 - \alpha)u^2]$$

is strictly convex, since $\mathcal{L}(x, u^1)$ and $\mathcal{L}(x, u^2)$ are strictly convex.] Since $u^1, u^2 \in D(\gamma)$, $T^k = \{x \mid \mathcal{L}(x, u^k) \leq \gamma(u^k)\}$ $(k = 1, 2)$ are single-point sets (hence bounded). Let $b_1 = \mathcal{L}[x^1, \alpha u^1 + (1 - \alpha)u^2]$, and $b_2 = \mathcal{L}[x^2, \alpha u^1 + (1 - \alpha)u^2]$. Let $T^3 = \{x \mid \mathcal{L}[x, \alpha u^1 + (1 - \alpha)u^2] \leq \min(b_1, b_2)\}$. (This is nonempty, since either x^1 or x^2 is contained in it.) Let $T^4 = \{x \mid \mathcal{L}(x, u^1) \leq b_1\}$, and $T^5 = \{x \mid \mathcal{L}(x, u^2) \leq b_2\}$. By Theorem 23, T^1 nonempty and bounded implies T^4 bounded, and T^2 bounded and nonempty implies T^5 bounded. If $x^0 \in T^3$, then $\mathcal{L}[x^0, \alpha u^1 + (1 - \alpha)u^2] = \alpha\mathcal{L}(x^0, u^1) + (1 - \alpha)\mathcal{L}(x^0, u^2) \leq \min(b_1, b_2)$. Clearly either $x^0 \in T^4$ or $x^0 \in T^5$. Thus $T^3 \subset T^4 \cup T^5$, and T^3 is a bounded set. Since the set of points that minimize the strictly convex function $\mathcal{L}[x, \alpha u^1 + (1 - \alpha)u^2]$ is in T^3, that set consists of one point. This completes the proof that the set of points $D(\gamma)$ is convex.

7.4 Dual Method for Strictly Convex Problems

Since $\gamma[\alpha u^1 + (1 - \alpha)u^2] = \inf_x \mathcal{L}[x, \alpha u^1 + (1 - \alpha)u^2] \geq \alpha \inf_x \mathcal{L}(x, u^1) + (1 - \alpha)\inf_x \mathcal{L}(x, u^2) = \alpha\gamma(u^1) + (1 - \alpha)\gamma(u^2)$, the auxiliary function γ is concave in u.

The optimality of u^* follows, since $\gamma(u) = \mathcal{L}[x(u), u] \leq v^*$ (from Theorem 22), and $\gamma(u^*) = v^*$ from Theorem 23. Q.E.D.

Application of the theorem is illustrated by the following example.

Example.

$$\text{minimize } x_1 + e^{-x_1}$$

subject to

$$x_1 - 1 \geq 0.$$

The solution is at $x_1^* = 1$, and the associated $u_1^* = 1 - 1/e$. It is easily verified that $\nabla^2 \mathcal{L}(x^*, u^*)$ is greater than zero. The Lagrangian gradient is $\nabla_x \mathcal{L}(x, u) = 1 - e^{-x_1} - u_1$. Setting this equal to zero and solving for x_1 in terms of u_1 yields

$$x_1(u_1) = -\ln(1 - u_1).$$

This is not meaningful for $u \geq 1$. Another way of seeing this is to notice that for $u \geq 1$ the Lagrangian $\mathcal{L}(x, u) = x_1 + e^{-x_1}$ is minimized at $x_1 = +\infty$; that is, no finite minimum exists.

For $0 \leq u < 1$, as indicated in the theorem, $x(u_1)$ is a single point. Substituting $x_1(u_1)$ in the dual yields

$$\text{maximize } \gamma(u_1) = [-\ln(1 - u_1)](1 - u_1) + 1$$

subject to

$$u_1 \geq 0.$$

(This can be verified as the maximization of a concave function.) The solution is unconstrained in u_1 and can be found by differentiating γ. This yields $u_1 = 1 - 1/e$, and the initial u_1^* is also the solution to the dual programming problem (β) as stated in the theorem.

A most important fact about Problem β is that in the interior of $D(\gamma)$ the auxiliary function $\gamma(u)$ is continuously differentiable.

Theorem 36 [Differentiability of $\gamma(u)$]. Under the hypotheses of Theorem 35, $\gamma(u)$ is continuously differentiable in the interior of $D(\gamma)$, and at any $u^0 \in \text{int } D(\gamma)$

$$\left.\frac{\partial \gamma}{\partial u_i}\right|_{u^0} = -g_i[x(u^0)], \quad i = 1, \ldots, m.$$

Furthermore, if $u_i^0 = 0$, then $\partial \gamma / \partial u_i^+ \big|_{u^0}$ exists and equals $-g_i[x(u^0)]$.

Other Unconstrained Minimization Techniques

Proof. Let $h > 0$, and e^i denote the ith unit vector. Then

$$\frac{\gamma(u^0 + he^i) - \gamma(u^0)}{h} \leq \frac{\mathcal{L}[x(u^0), u^0 + he^i] - \mathcal{L}[x(u^0), u^0]}{h}$$

$$= -g_i[x(u^0)].$$

On the other hand,

$$\frac{\gamma(u^0 + he^i) - \gamma(u^0)}{h} \geq \frac{\mathcal{L}[x(u^0 + he^i), u^0 + he^i] - \mathcal{L}[x(u^0 + he^i), u^0]}{h}$$

$$= -g_i[x(u^0 + he^i)].$$

Since g_i is continuous in x and x is continuous in u (the last statement has not been proved but follows from the strict convexity), taking the limit as $h \to 0$ yields the proof that

$$\left.\frac{\partial \gamma}{\partial u_i^+}\right|_{u^0} = -g_i[x(u^0)].$$

The previous remarks hold when $u_i^0 \geq 0$. When $u_i^0 > 0$, a similar proof yields $\partial \gamma / \partial u_i^- \,|\, u^0 = -g_i[x(u^0)]$, and the proof is complete. Q.E.D.

In order to solve Problem β as a convex programming problem in u, any technique can be used that utilizes first partial derivatives. One algorithm, which is a modification of the method of steepest ascent (see Section 8.2), proceeds as follows.

1. Let $u^0 \geq 0$ be a starting point of the process. Assume $u^0 \in D(\gamma)$.
2. If u^k is a set of multipliers for which the Lagrangian has been minimized, and if $x^k = x(u^k)$ is the unique minimizing x, then compute a *direction vector* s^k as

$$s_i^k = 0 \quad \text{if} \quad u_i^k = 0 \quad \text{and} \quad -g_i(x^k) < 0; \tag{7.15}$$

$$s_i^k = -g_i(x^k) \quad \text{otherwise.} \tag{7.16}$$

Find α^k such that $\gamma[u^k + \alpha^k s^k] = \max_{\alpha \in S^k} \gamma[u^k + \alpha s^k]$, where $S^k = \{\alpha \,|\, (u^k + \alpha s^k) \geq 0\}$. This involves a search in a single variable, which may end if some component of $u^k + \alpha s^k$ becomes zero or if the unconstrained maximum of γ is found in the direction s^k.

4. Repeat 2 and 3 until s^k, as given by (7.15) and (7.16), is acceptably close to zero.

To compute $\gamma(u)$ it is necessary to find the x that minimizes the Lagrangian. For certain problems, for example, separable programming problems, this is not very difficult.

7.4 Dual Method for Strictly Convex Problems

A separable programming problem is one that can be written

$$\text{minimize } f(x) = \sum_{j=1}^{n} f_j(x_j)$$

subject to

$$g_i(x) = \sum_{j=1}^{n} g_{ij}(x_j) \geq 0, \quad i = 1, \ldots, m.$$

Then, for any u^0,

$$\inf_x \mathcal{L}(x, u^0) = \sum_{j=1}^{n} \inf_{x_j} \left[f_j(x_j) - \sum_{i=1}^{m} u_i^0 g_{ij}(x_j) \right].$$

Thus a one-dimensional search in each of n components is all that is required to minimize the Lagrangian. (See Section 8.5 for a discussion of one-variable search methods.)

For problems where each x_j is a *vector* rather than a single variable, the advantages of the separable form of the Lagrangian can also be used.

Other methods for solving Problem β [$\max_{u \in D(\gamma)} \gamma(u)$] use first partial derivatives. In particular, success has been reported using the method of Fletcher and Reeves [57]. Presumably any method of conjugate directions (see Section 8.2) would do as well.

The strictly convex dual method presented above can be generalized to handle problems where the minimization of the Lagrangian can be restricted to some subset C of E^n. Thus it is possible to include some of the constraints (for example, $x_j \geq 0$, $j = 1, \ldots, n$) in C rather than including them explicitly in the Lagrangian. The assumption of strict convexity is still strong enough to ensure convergence, although the proofs would have to be modified slightly. The reader is referred to [41] for a fuller development along these lines.

Application of the Dual Method to General Convex Programming Problems

As stated in the preceding discussion, the dual method described can solve only strictly convex problems. For a general convex problem there is no guarantee that, for any u^0, the Lagrangian function has a finite unconstrained minimum. In particular, it is obvious that for a linear programming problem there is only one set of choices of u, namely, the optimal $u = u^*$, for which the Lagrangian is constant and hence, trivially, has finite unconstrained minima. In the following development there is defined a very simple sequence of ϵ-perturbations of the original convex programming problem, which can be solved by the dual approach. As the perturbed problem approaches the original problem, its solution approaches the original solution.

136 Other Unconstrained Minimization Techniques

Consider the following convex programming problem:

Definition.
$$\text{minimize } f(x) + \epsilon^k \|x\|^2 \qquad C(\epsilon^k)$$
subject to
$$g_i(x) \geq 0, \quad i = 1, \ldots, m,$$
where $\epsilon^k > 0$.

The ϵ-perturbation approach is to apply the dual method to the sequence of problems $\{C(\epsilon^k)\}$ for a sequence $\{\epsilon^k\}$ approaching zero.

Corresponding to each $[C(\epsilon^k)]$ is the dual programming problem:

$$\text{maximize } \gamma^k(u) = \mathcal{L}^k[x(u), u] = f[x(u)] + \epsilon^k \|x(u)\|^2 - \sum_{i=1}^{m} u_i g_i[x(u)] \quad \beta(\epsilon^k)$$
subject to
$$u \in D^k(\gamma)$$

[which has the definition analogous to $D(\gamma)$].

The following theorem guarantees the convergence of solutions of $\{\beta(\epsilon^k)\}$ and $\{C(\epsilon^k)\}$ to a solution of (C).

Theorem 37 [Convergence of ϵ-Lagrange Multiplier Approach to Solution of Convex Programming Problem]. If $f, -g_1, \ldots, -g_m$ are convex functions, and if $R \equiv \{x \mid g_i(x) \geq 0, i = 1, \ldots, m\}$ is nonempty, then there is a unique finite solution x^k to every perturbed problem $[C(\epsilon^k)]$. Furthermore, with respect to $\beta(\epsilon^k)$, $D^k(\gamma)$ is $(E^m)^+$; that is, the Lagrangian $\mathcal{L}^k(x, u^0)$ has a finite minimum for every $u^0 \geq 0$. If, in addition, for Problem $[C(\epsilon^k)]$ there is a finite u^k associated with x^k satisfying (6.10–6.13), then u^k is a solution to $\beta(\epsilon^k)$. Finally, if there is at least one finite point solving (C), then the sequence $\{x^k\}$ has as its unique limit point $\bar{x}$, the solution of (C) with minimum norm.

Proof. We first prove a preliminary result. If $s(x)$ is any convex function, then for any M, every $\epsilon^k > 0$, $\{x \mid s(x) + \epsilon^k \|x\|^2 \leq M\}$ is a bounded set. Let $\{y^l\}$ be an infinite sequence of points where $\lim_{l \to \infty} \|y^l\| = +\infty$. Let y^0 be any point. We construct a sphere of radius δ about y^0 and denote B as the boundary of that sphere. We consider only points y^l far enough along in the infinite unbounded sequence so that they are outside the sphere. For each y^l let $x_B{}^l$ be the point on the line joining y^0 and y^l that is on B. Let λ^l be the scalar, where $0 < \lambda^l < 1$ and

$$\lambda^l y^l + (1 - \lambda^l) y^0 = x_B{}^l.$$

Then

$$s(y^l) + \epsilon^k \|y^l\|^2 \geq \frac{s(x_B{}^l) - s(y^0)}{\lambda^l} + s(y^0) + \epsilon^k \|y^l\|^2$$

7.5 Lagrange Multiplier Technique for Resource Allocation Problems

(convexity of s)

$$\geq \frac{\bar{v}_B - s^0}{\lambda^l} + s(y^0) + \epsilon^k \|y^l\|^2,$$

(where $\bar{v}_B = \min_B s(x)$)

$$\geq \frac{\bar{v}_B - s^0}{\lambda^l} + s(y^0) + \epsilon^k \left[\frac{\delta^2}{(\lambda^l)^2} + \frac{2(y^0)^T(x_B^{\,l} - y^0)}{\lambda^l} + \|y^0\|^2 \right]$$

(definition of λ^l, and the sphere about y^0). Clearly then, since

$$\lambda^l \to 0(\|y^l\| \to +\infty),$$

the term with $(\lambda^l)^2$ in the denominator dominates, and $s(y^l) + \epsilon^k \|y^l\|^2$ must go to $+\infty$. This proves our preliminary result.

Letting $s(x) = f(x)$ proves that a finite solution exists for every $[C(\epsilon^k)]$. Uniqueness follows from the strict convexity of $f(x) + \epsilon^k \|x\|^2$. Letting $s(x) = f(x) - \sum_{i=1}^{m} u_i^0 g_i(x)$ proves that $D^k(\gamma) = (E^m)^+$.

From the assumption of a u^k associated with x^k satisfying the necessary and sufficient conditions for x^k to be a solution of Problem $C(\epsilon^k)$, it follows trivially that u^k solves $\beta(\epsilon^k)$.

The final part of the theorem is proved as follows. Consider the sequence $\{x^k\}$, where each x^k is the unique solution to $[C(\epsilon^k)]$. Let $\bar{x}$ denote the solution to (C) with minimum norm. (We have assumed at least one finite solution; hence the solution with minimum norm is well defined.) Clearly, since each x^k is feasible,

$$f(\bar{x}) \leq f(x^k), \quad \text{all } k.$$

If any limit point of $\{x^k\}$ is unbounded in norm, eventually $\epsilon^k \|\bar{x}\|^2 < \epsilon^k \|x^k\|^2$, which means that x^k is not a solution to $[C(\epsilon^k)]$. Thus all limit points of $\{x^k\}$ are finite. From this we conclude $\lim_{k\to\infty} \epsilon^k \|x^k\|^2 = 0$. Hence $\lim_{k\to\infty} f(x^k) = f(\bar{x}) = \inf_R f(x)$; otherwise eventually

$$f(\bar{x}) + \epsilon^k \|\bar{x}\|^2 < f(x^k) + \epsilon^k \|x^k\|^2,$$

since the second terms in the inequality both go to zero. Finally, since $f(\bar{x}) \leq f(x^k)$ and every limit point of $\{x^k\}$ solves (C), for any other distinct limit point eventually $\|x^k\|^2 > \|\bar{x}\|^2$ (for those k where x^k approaches the other limit point), and this contradicts the assertion that x^k solves $[C(\epsilon^k)]$.
Q.E.D

7.5 GENERALIZED LAGRANGE MULTIPLIER TECHNIQUE FOR RESOURCE ALLOCATION PROBLEMS

Another technique for nonlinear programming problems that uses Lagrange multipliers was first presented in [40]. In contrast to the method of Section

7.4, this technique is intended to apply to a wider class of problems. In particular, the space of feasible points need not be dense subsets of E^n, but may consist of discrete point sets. The functions defining the objective and constraints need not be continuous. The motivation behind the generalized Lagrange multiplier technique (GLMT) is that the dual variables have a "meaning" in optimal resource allocation problems that is independent of their algebraic derivation (as exemplified in Section 2.1) for differentiable functions [see (2.48), Section 2.4].

In the original presentation the motivation is provided by viewing the programming problem as that of determining non-negative activity levels $(x_1, \ldots, x_n)$ that

$$\text{maximize } F(x_1, \ldots, x_n) \tag{7.17}$$

subject to constraints

$$G_i(x_1, \ldots, x_n) \leq b_i, \quad i = 1, \ldots, m, \tag{7.18}$$

where each $G_i(x_1, \ldots, x_n)$ represents the amount of resource i consumed by activity levels $(x_1, \ldots, x_n)$ and b_i is the amount of resource i available. The vector x is further restricted to a set X, called a strategy set. Typically the non-negativity requirements, restriction to integer values, and so on, are imposed by definition of the set X. Thus we have the additional constraint

$$x \in X \tag{7.19}$$

The GLMT proceeds as follows.

1. Select values u_i^0, $i = 1, \ldots, m$.
2. Find an $x(u^0)$ that gives

$$\sup_{x \in X} \left[F(x) - \sum_{i=1}^{m} u_i^0 G_i(x) \right]. \tag{7.20}$$

3. The remainder of the algorithm consists of changing u until the resources associated with the nonzero u_i are "fully utilized" at the current activity level.

Let x^k be the activity level that solves

$$\sup_{x \in X} \left[F(x) - \sum_{i=1}^{m} u_i^k G_i(x) \right]. \tag{7.21}$$

If for all i, where $u_i^k = 0$,

$$G_i(x^k) \leq b_i, \tag{7.22}$$

and for all i, where $u_i^k > 0$,

$$G_i(x^k) = b_i, \tag{7.23}$$

the algorithm is terminated.

7.5 Lagrange Multiplier Technique for Resource Allocation Problems

One algorithm† that has had computational success for adjusting the u_i's is as follows. Let $\epsilon_1, \ldots, \epsilon_m$ be a set of strictly positive values, one for each resource. Let j be the index of a resource where (7.22) does not hold. Then u_j^{k+1} is altered according to the following possibilities.

When $G_j(x^k) > b_j$, and

(i) $G_j(x^k) > b_j \geq G_j(x^{k-1})$, $\quad \epsilon_j^{k+1} = \dfrac{\epsilon_j^k}{4}, \quad u_j^{k+1} = (1 + \epsilon_j^{k+1})u_j^k;$

(ii) $G_j(x^{k-1}) > G_j(x^k) > b_j$, $\quad \epsilon_j^{k+1} = 1.3\epsilon_j^k, \quad u_j^{k+1} = (1 + \epsilon_j^{k+1})u_j^k;$

(iii) $G_j(x^k) \geq G_j(x^{k-1}) > b_j$, $\quad \epsilon_j^{k+1} = 2\epsilon_j^k, \quad u_j^{k+1} = (1 + \epsilon_j^{k+1})u_j^k.$

When $G_j(x_j) < b_j$, and

(iv) $b_j > G_j(x^k) > G_j(x^{k-1})$, $\quad \epsilon_j^{k+1} = 1.3\epsilon_j^k, \quad u_j^{k+1} = (1 - \epsilon_j^{k+1})u_j^k;$

(v) $b_j > G_j(x^{k-1}) \geq G_j(x^k)$, $\quad \epsilon_j^{k+1} = 2\epsilon_j^k, \quad u_j^{k+1} = (1 - \epsilon_j^{k+1})u_j^k;$

(vi) $G_j(x^{k-1}) \geq b_j > G_j(x^k)$, $\quad \epsilon_j^{k+1} = \dfrac{\epsilon_j^k}{4}, \quad u_j^{k+1} = (1 - \epsilon_j^{k+1})u_j^k.$

When

(vii) $G_j(x^k) = b_j$, $\quad \epsilon_j^{k+1} = \dfrac{\epsilon_j^k}{4}, \quad u_j^{k+1} = u_j^k.$

The motivation for ceasing the iterations when all resources with nonzero Lagrange multipliers are approximately fully utilized is provided in the following simple yet powerful theorem (called in [40], page 401, the Main Theorem).

Theorem 38 [Optimality of Lagrangian Maximum for Corresponding Constrained Problem]. If for some non-negative multipliers $\{u_i^0\}$, $x(u^0)$ solves
$$\max_{x \in X} [F(x) - \sum u_i^0 G_i(x)],$$
then $x(u^0)$ also solves
$$\text{maximize } F(x)$$
subject to
$$G_i(x) \leq G_i[x(u^0)] \quad \text{(for all } i, \text{ where } u_i^0 > 0) \tag{7.24}$$
and
$$x \in X. \tag{7.25}$$

† This algorithm is due to Hugh D. Everett III.

Proof. Let y be any other point in X satisfying (7.24) and (7.25). Then

$$F[x(u^0)] \geq F[x(u^0)] - \sum u_i^0 \{G_i[x(u^0)] - G_i(y)\}$$
$$\geq F(y) - \sum u_i^0 G_i(y) + \sum u_i^0 G_i(y)$$
$$= F(y). \qquad \text{Q.E.D.}$$

It follows from Theorem 38 that if some x^k satisfies (7.22) and (7.23), that x^k is a solution to the original programming problem given by (7.17–7.19). [For those resources i where $u_i^k = 0$, (7.22) guarantees feasibility.] A simple modification of Theorem 38 then proves the optimality of x^k.

At Step 3 the aim is to modify the multipliers so that the convergence criteria (7.22) and (7.23) are satisfied. Thus if for some i, $u_i^k > 0$ and $G_i(x^k) \neq b_i$, then u_i^k should be adjusted to make equality hold. A criterion for effecting Step 3 is suggested by the following theorem (called in [40], page 405, the Lambda Theorem).

Theorem 39 [Bounds on Ratio of Increased Payoff to Increased Resource Allocation Using Multipliers]. Let $\{u_i^1\}$ and $\{u_i^2\}$ be two sets of Lagrange multipliers whose corresponding Lagrangian maxima are given by x^1 and x^2. Suppose that they have the property that $G_j(x^1) > G_j(x^2)$ for some index j, and $G_i(x^1) = G_i(x^2)$ for all $i \neq j$ (if both u_i^1 and u_i^2 are equal to zero, equality need not hold). Then

$$u_j^2 \geq \frac{F(x_1) - F(x^2)}{G_j(x^1) - G_j(x^2)} \geq u_j^1. \qquad (7.26)$$

Proof. By assumption,

$$F(x^1) - \sum u_i^1 G_i(x^1) \geq F(x^2) - \sum u_i^1 G_i(x^2).$$

Combining, and using the fact that $G_i(x^1) = G_i(x^2)$ for all $i \neq j$, where both u_i^1 and u_i^2 are not zero and $G_j(x^1) > G_j(x^2)$, then

$$\frac{F(x^1) - F(x^2)}{G_j(x^1) - G_j(x^2)} \geq u_j^1.$$

Also, by assumption,

$$F(x^2) - \sum u_i^2 G_i(x^2) \geq F(x^1) - \sum u_i^2 G_i(x^1).$$

Proceeding as above, the other inequality is obtained. Q.E.D.

Then (7.26) implies that in order to use *more* of resource j and maintain the levels of the other resources (with nonzero u_i^k's) its associated current multiplier value u_j^k should be *decreased*. The converse statement holds if the jth resource level is to be increased.

7.5 Lagrange Multiplier Technique for Resource Allocation Problems 141

Another way of looking at this way of adjusting u_j^k is that if the appropriate differentiability assumptions hold and u^k is fixed, then u_j^k is the partial derivative of the optimum value of $F(x)$ [in the programming problem given by (7.24) and (7.25)] with respect to the jth resource level. In economic terms, u_j^k is the rate of increase in production that would result if one more resource unit of type j were available and were optimally allocated among the activity levels.

One would expect that by decreasing the "shadow price" u_j^k, more of resource j would be used, and by increasing it, less would be used.

Having described and somewhat motivated the GLMT, we turn briefly to a comparison with the method treated in Section 7.4.

Both methods adjust the multipliers $\{u_i\}$ and then optimize the Lagrangian function. If we make the correspondence $f(x) = -F(x)$ and $g_i(x) = -G_i(x) + b_i(i = 1, \ldots, m)$, it is clear that the minimization of Section 7.4 is the same as the maximization of this section. For the dual approach, $x \in E^n$, and for the GLMT $x \in X$. In [43] a fuller development of the dual approach of Section 7.4 is contained, and there x may be restricted to a set C, which differs from the generality of X in that C must be a dense subset of E^n whereas X can be *any* subset of E^n (such as the lattice points of integers) over which the Lagrangian function assumes its maximum.

More interesting is the similarity between the two methods and how they change the multipliers. In the modified method of steepest ascent of Section 7.4, the *direction* vector in which the u_i's are changed is

$$s_i = 0 \quad \text{if} \quad u_i^k = 0 \quad \text{and} \quad -g_i(x^k) < 0; \tag{7.27}$$

$$s_i = -g_i(x^k) \quad \text{otherwise.} \tag{7.28}$$

As concerns the *magnitude* of the change, since the aim of the strict dual method is to maximize the Lagrangian as a function of u, subject to the nonnegativity restrictions, the scalar α^k, where $u^{k+1} = u^k + \alpha^k s^k$, is well defined.

In the GLMT each u_i^k is changed in direction proportional to $-g_i(x) = +G_i(x) - b_i$ (for those u_i^k's not equal to zero). The factor of proportionality depends on the rate at which the ith allocation changed from iteration $(k-1)$ to iteration k. If the change from u_i^{k-1} to u_i^k resulted in a "crossover" (see Rules i and vi), then it is assumed that u^k would be approximately equal to its optimum value and that the proportionality factor should be decreased.

When the change in the multiplier results in an inferior allocation, then the proportionality factor is increased at a faster rate, and the multiplier changed accordingly [(iii) and (v)].

The following described the kinds of problems on which the GLMT appears to work best.

Cell Problems

We assume that the objective function $F(x)$ can be divided into l independent "cells," which separately contribute to the over-all payoff. That is,

$$F(x) = \sum_{c=1}^{l} F_c(y_c),$$

where each y_c is a partial vector of x's (activity levels) that contribute to the payoff of the cth cell. Let X_c be the strategy space of y_c. We further assume that each constraint is separable with respect to the y_c, so that

$$G_i(x) = \sum_{c=1}^{l} G_i^c(y_c) \leq b_i, \qquad i = 1, \ldots, m.$$

Then the Lagrangian maximization is

$$\sum_{c=1}^{l} \max_{y_c \in X_c} \left[F_c(y_c) - \sum_{i=1}^{m} u_i^0 G_i^c(y_c) \right].$$

Thus a very important fact results, namely, that the total maximization of the Lagrangian can be achieved by separately maximizing over the activities contributing to each cell.

In some very large cell problems the resources i are consumed *independently* of each other; that is, no variables appearing in constraint i appear in constraint j. The variables x are best described as a matrix array $\{x_{ic}\}$, where $i = 1, \ldots, m$, and $c = 1, \ldots, l$. Thus x_{ic} means the amount of resource i allocated to cell c. Then the payoff function can be written $\sum_{c=1}^{l} f_c(y_c)$, where $y_c = (x_{1c}, \ldots, x_{mc})$. Clearly, consumption of the ith resource is independent of consumption of the jth resource when the variables are of this form, the problem being even more uncoupled than before. Each constraint has the form

$$G_i = \sum_{c=1}^{l} r_c^i(x_{ic}) \leq b_i.$$

The Lagrangian maximization is then

$$\sum_{c=1}^{l} \max_{y_c \in X_c} \left\{ f(x_{1c}, \ldots, x_{mc}) - \sum_{i=1}^{m} u_i^0 r_c^i(x_{ic}) \right\},$$

where $y_c = (x_{1c}, \ldots, x_{mc})$. Because each u_i^0 multiplies only those variables appearing in the ith constraint, one would expect the effect of changing any u_i in Step 3 to affect *mainly* the allocation of resource i; that is, the Lagrangian is virtually separable as a function of u.

7.6 Solution by a Single Unconstrained Minimization

The rules for changing u^k may be regarded as a heuristic guide for finding the maximum of an unconstrained separable function by adjusting the partial derivatives [given by $g_i = -G_i(x) + b_i$] by a weighting factor that is itself affected by the changes in the derivative from iteration to iteration.

As a tool for solving large-scale resource allocation problems with cell structure, the generalized Lagrange multiplier technique should prove quite useful. A lack of a precise statement of conditions on the programming problems that guarantee its convergence, the lack of published experience with the method, and its theoretical difficulties with certain problems keep it in the realm of conjectural rather than proved algorithms.

7.6 SOLUTION OF CONSTRAINED PROBLEM BY A SINGLE UNCONSTRAINED MINIMIZATION

In this section is presented an exterior point algorithm, which, for appropriate choices of the parameter t, has an unconstrained minimum that is also a constrained minimum of the convex programming problem (C). The method was proposed and proved by Zangwill [115, 117] and is similar to the suggestion of Ablow and Brigham [2].

The exterior penalty function is

$$S(x, t) = f(x) - t \sum_{i=1}^{m} \min \left[g_i(x), 0 \right]. \tag{7.29}$$

Clearly this satisfies the requirements of an exterior point function of Section 4.1, and is convex for the convex programming problem (C).

That an unconstrained minimum to (7.29) for large t is also a constrained solution to (C) is proved in the following theorem.

Theorem 40 [Existence of t^0 for Which Unconstrained Minimum Is Also a Constrained Minimum]. If (a) $f, -g_1, \ldots, -g_m$ are convex functions, (b) $R^0 = \{x \mid g_i(x) > 0, i = 1, \ldots, m\}$ is nonempty, and (c) there exists a finite point x^* solving the convex programming problem (C), then there exists a value $t^0 > 0$ such that for $t \geq t^0$ any unconstrained minimum of the exterior penalty function (7.29) is feasible and thus solves the constrained problem (C).

Proof. Let $x^0 \in R^0$. Define

$$a = \min_i \{g_i(x^0)\}. \tag{7.30}$$

Let

$$b = f(x^0) - f(x^*), \tag{7.31}$$

where x^* is an optimal point assured by assumption c. Let $t^0 = (b+1)/a$. We show that for any infeasible point w there exists a feasible point v such that

$$S(w, t^0) > S(v, t^0).$$

Since $S(x^*, t^0) \leq S(v, t^0)$ for any $v \in R$, it follows that any solution of (C) is an unconstrained minimum of $S(x, t^0)$. Clearly, for $t \geq t_0$, the same conclusion holds.

The point v is defined as follows. Let v be the unique point on the line connecting w and x^0, which is on the boundary of the feasible region. Let $A = \{i \mid g_i(v) = 0\}$, so that for $i \notin A$, $g_i(v) > 0$.

We define an auxiliary function

$$s(x) = f(x) - t^0 \sum_{i \in A} g_i(x).$$

Then

$$s(x^0) = f(x^0) - \frac{b+1}{a} \sum_{i \in A} g_i(x^0)$$

$$\leq f(x^0) - \frac{b+1}{a} \left[\min_i g_i(x^0) \right]$$

$$= f(x^*) - 1 \quad \text{[from (7.30) and (7.31)]}$$

$$< f(v) \quad \text{(since } v \text{ is primal feasible and } x^* \text{ is optimal)}$$

$$= s(v) \quad \left[\text{by definition } \sum_{i \in A} g_i(v) = 0 \right].$$

Thus

$$s(x^0) < s(v). \qquad (7.32)$$

The point v is on the interior of the line segment connecting w and x^0, so that there is a λ, $0 < \lambda < 1$ such that $v = \lambda x^0 + (1-\lambda)w$. Because $s(x)$ is convex,

$$s(v) \leq \lambda s(x^0) + (1-\lambda)s(w)$$

$$< \lambda s(v) + (1-\lambda)s(w) \quad \text{[by (7.32)]}.$$

Transposing and dividing yields

$$s(v) < s(w). \qquad (7.33)$$

Since $s(v) = f(v)$ and $f(v) = S(v, t^0)$ (v being feasible),

$$s(v) = S(v, t^0). \qquad (7.34)$$

By concavity,

$$g_i(v) \geq \lambda g_i(x^0) + (1-\lambda)g_i(w),$$

so that for all $i \in A$, $g_i(w) \leq [g_i(v) - \lambda g_i(x^0)]/(1 - \lambda) < 0$. Thus

$$\begin{aligned} s(w) &= f(w) - t^0 \sum_{i \in A} g_i(w) \\ &= f(w) - t^0 \sum_{i \in A} \min[g_i(w), 0] \\ &\leq f(w) - t^0 \sum_{i=1}^{m} \min[g_i(w), 0] \\ &= S(w, t^0). \end{aligned} \qquad (7.35)$$

Then, using (7.34), (7.33), and (7.35),

$$S(v, t^0) = s(v) < s(w) \leq S(w, t^0). \qquad \text{Q.E.D.}$$

Theorem 40 is more of theoretical value than any practical use since the function (7.29) is not *differentiable* at any boundary point of R. Since most techniques for minimizing unconstrained functions (see Section 8.2) rely on first partial derivatives, this has serious computational implications. However, it *is* a convex function (and hence continuous), and perhaps some of the techniques that use only function values to find unconstrained minima can be used on it successfully.

One is tempted to say that the minimization of (7.29) decomposes into a series of univariate minimizations for *separable* problems. However, the function is only separable over regions of E^n, and not over the whole space. Minimizing separately in each component can lead (and usually does) to positions that are not minima but from which univariate moves result in function increases.

A small example illustrating Theorem 40 is given below.

Example.
$$\text{minimize } x_1^2 - 2x_1$$
subject to
$$-x_1 \geq 0.$$

The solution is at $x_1^* = 0$. For each t the S function takes two different forms, one for the region where $x_1 > 0$, and one for $x_1 \leq 0$. For $x_1 > 0$, $S(x_1, t) = x_1^2 + (t - 2)x_1$, and for $x_1 \leq 0$, $S(x_1, t) = x_1^2 - 2x_1$. For $t < 2$, the unconstrained minimum is at $(2 - t)/2$. For $t \geq 2$, the unconstrained minimum agrees with the constrained minimum and is at $x_1 = 0$.

7.7 COMBINED UNCONSTRAINED-SIMPLICIAL ALGORITHMS

The motivation behind the use of interior and exterior point unconstrained minimization algorithms for converting constrained optimization problems

146   Other Unconstrained Minimization Techniques

to sequences of unconstrained problems was primarily to avoid the difficulties imposed by *nonlinear* constraints. Movement in a nonlinear hyperspace is much more difficult than movement in a linear hyperplane. It seems reasonable, in fact, that a modification of the basic unconstrained algorithms that treats the linear constraints (particularly upper or lower bounds on the variables if they are present) separately would be more efficient. The modification proposed here is to include only the nonlinear constraints in the auxiliary function whose minimum is to be determined, subject to the satisfaction of the linear inequality constraints.

The mixed nonlinear-linear programming problem is written as follows:

$$\text{minimize } f(x) \tag{NL}$$

subject to

$$g_i(x) \geq 0, \quad i = 1, \ldots, m, \tag{7.36}$$

$$Ax - b \geq 0. \tag{7.37}$$

A revision of the basic interior minimization algorithm of Section 3.2 that applies to Problem NL is as follows.

Let $R^0 = \{x \mid g_i(x) > 0, i = 1, \ldots, m\}$, and $H = \{x \mid Ax - b \geq 0\}$. Let $I(g)$ be an interior point unconstrained minimization function, that is, $I(g)$ is continuous in R^0, and $\lim_{k \to \infty} I(g^k) = +\infty$ for any infinite sequence $\{g^k\}$, $g^k > 0$, where every limit point of $\{x^k\}$ is a boundary point of R^0.

1. Define the function

$$U(x, r_1) = f(x) + r_1 I[g(x)],$$

where $r_1 > 0$. As a starting point determine an $x^0 \in R^0 \cap H$.

2. Proceed from x^0 to a point $x(r_1) = x^1$, which is a local minimum of $U(x, r_1)$ in $R \cap H$. Presumably x^1 will be unconstrained relative to R in that it will lie in $R^0 \cap H$.

3. Starting from x^1, find a local minimum of $U(x, r_2)$, where $r_1 > r_2 > 0$, in $R \cap H$.

4. Continuing in this fashion, find a local minimum of $U(x, r_k)$, starting from x^{k-1}, for a strictly decreasing positive null sequence $\{r_k\}$.

Under appropriate assumptions this algorithm generates points that approach local minima of Problem NL. In fact, under the same assumptions as those of Theorem 8, there exist for values of r_k small enough, minima of U in $R \cap H$ converging to local isolated compact sets of minima of f in $R \cap H$. Obvious modifications of the proof of that theorem give this result.

When Problem NL is a convex programming problem and the I function is chosen as in Section 6.3, then the revised algorithm involves a sequence of minimizations of a convex function subject to linear inequality constraints.

7.7 Combined Unconstrained-Simplicial Algorithms

An important feature of the revised algorithm is the preservation of the dual properties of interior point algorithms for convex programming problems. That is, at the solution of each linearly constrained problem, a dual feasible point is generated with the usual resulting lower bound on the optimum value v^*.

Thus if

$$U(x, r_k) = f(x) + r_k \sum_{i=1}^{m} I_i[g_i(x)],$$

where each I_i is a *differentiable, decreasing convex* function of g_i when $g_i > 0$, the dual variables at the minimum for the nonlinear inequality constraints (7.36) are

$$u_i^k = \frac{-r_k \, \partial I_i[g_i(x^k)]}{\partial g_i}, \qquad i = 1, \ldots, m. \tag{7.38}$$

The dual variables for the linear constraints (7.37) may or may not be explicitly available, depending on the algorithm used to solve the linearly constrained problem. In any event their existence is a consequence of the necessary conditions of Corollary 1. (The first-order constraint qualification holds for a linearly constrained problem, and thus Corollary 1 applies to every subproblem at Step 4.) Because of the complementarity condition (2.7), their contribution to the dual value is zero. Thus the lower bound available at every dual feasible point is

$$f(x^k) - \sum_{i=1}^{m} -r_k \frac{\partial I_i^k}{\partial g_i} g_i^k \leq v^*. \tag{7.39}$$

A more troublesome question is: Under what circumstances can the extrapolation techniques of Chapter 5 be applied to unconstrained simplicial algorithms?

Assume that the problem functions are twice continuously differentiable and that the sufficient conditions (Theorem 4) hold at a point x^* such that x^* is a local constrained minimum of Problem NL. Then there exists (u^*, t^*) such that (x^*, u^*, t^*) satisfies

$$\begin{aligned}
g_i(x) &\geq 0, & i &= 1, \ldots, m, \\
Ax - b &\geq 0, & & \\
u_i g_i(x) &= 0, & i &= 1, \ldots, m, \\
u_i &\geq 0, & i &= 1, \ldots, m,
\end{aligned} \tag{7.40}$$

$$\begin{aligned}
t_j(a_j x - b_j) &= 0, & j &= 1, \ldots, l, \\
t_j &\geq 0, & j &= 1, \ldots, l,
\end{aligned} \tag{7.41}$$

$$\nabla f - \sum_{i=1}^{m} u_i \nabla g_i - \sum_{j=1}^{l} t_j a_j^T = 0. \tag{7.42}$$

148   Other Unconstrained Minimization Techniques

Assume strict complementarity holds for (7.40) and (7.41). Then for every nonzero vector y such that

$$y^T \nabla g_i^* = 0 \quad \text{for all} \quad i \in D^* = \{i \mid u_i^* > 0\}$$

and such that

$$a_j y = 0 \quad \text{for all} \quad j \in J^* = \{j \mid t_j^* > 0\}$$

it follows that

$$y^T \left[\nabla^2 f^* - \sum_{i=1}^m u_i \nabla^2 g_i^* \right] y > 0.$$

(Here a_j is a $1 \times n$ vector representing the jth row of A.) Assume further that the gradients of the binding constraints at x^* are linearly independent.

The Jacobian matrix of (7.40–7.42) with respect to (x, u, t) [at (x^*, u^*, t^*)] is

$$\begin{bmatrix} \nabla^2 f^* - \sum_{i=1}^m u_i^* \nabla^2 g_i^*, & -\nabla g_1^*, \ldots, -\nabla g_m^*, & -A^T \\ u_1^* \nabla^T g_1^* & g_1^* & \\ \cdot & \cdot & \\ \cdot & \cdot & \\ \cdot & \cdot & \\ u_m^* \nabla g_m^* & g_m^* & \\ t_1^* a_1 & & (a_1 x^* - b_1) \\ \cdot & & \cdot \\ \cdot & & \cdot \\ t_l^* a_l & & (a_l x^* - b_l). \end{bmatrix}$$

(7.43)

Under the assumptions we have made, it can be shown (as in Theorem 14) that (7.43) has an inverse. Then a theorem analogous to Theorem 14 can be proved that guarantees the existence of a unique *differentiable* trajectory $x(r)$ (for r small enough), every point of which is a local minimum of U in $R^0 \cap H$, that converges to x^*.

If further assumptions about the analyticity of the functions is made, a theorem analogous to Corollary 19 can be proved.

It is important to realize that the trajectory $x(r)$, even if it is unique, may not be differentiable everywhere. One would expect the trajectory to have nondifferentiable points that correspond to values of r where a linear constraint changes from equality satisfaction to a strict inequality satisfaction.

Thus extrapolation might give erroneous results. For r small enough, however, the preceding development assures us that the trajectory $x(r)$ will remain in a linear subspace identical with the one in which the local solution x^* lies.

A discussion of exterior point techniques as combined with these algorithms is not given here, as the development is analogous.

In the next section one method for solving nonlinear programming problems with linear constraints is combined with unconstrained interior techniques to illustrate the potentialities of the algorithms.

There are many other methods that handle linear constraints efficiently and can be combined with unconstrained algorithms. In particular, the "reduced gradient" method [45, 112] and Zoutendijk's "methods of feasible directions" [118, 119] are amenable to such a combined algorithm.

7.8 GRADIENT PROJECTION METHOD

One of the earliest methods for solving problems involving the minimization of a nonlinear function subject to linear constraints was suggested by J. B. Rosen [98] and is called the *gradient projection method* (GP). Certain characteristics of the method are (a) movement along the boundary of the polyhedron describing the feasible region, (b) the generation of a sequence of decreasing objective function values, and (c) the generation in the limit of the dual variables (generalized Lagrange multipliers).

The problem to which it is addressed is written

$$\text{minimize } \varphi(x)$$
subject to
$$Ax - b \geq 0.$$

Briefly stated, the gradient projection algorithm is as follows. Suppose that x^k is the current point. Let N^k represent a $q_k \times n$ ($q_k < n$) matrix of the transposed gradients of the linear constraints that are both currently satisfied and required to remain satisfied at the current iteration. (These gradients are assumed to be linearly independent for our purposes.) A matrix P^k (called the projection matrix) is formed as

$$P^k = I - (N^k)^T[N^k(N^k)^T]^{-1}N^k. \tag{7.44}$$

This matrix projects any vector into the subspace satisfying the constraints $N^k y = 0$. This is easily seen, since $N^k P^k y = 0 \cdot y = 0$. Every vector z in the space where $N^k z = 0$ is projected into itself, since $P^k z = z - 0 = z$.

150 Other Unconstrained Minimization Techniques

In particular P^k is used to project the gradient of φ at x^k. Then x^{k+1} is given by

$$x^{k+1} = x^k - \lambda^k P^k \nabla \varphi^k,$$

where λ^k is a scalar chosen to minimize φ along the vector $P^k \nabla \varphi^k$ without violating any currently satisfied constraints. If a new boundary is pierced before φ is minimized, its gradient is added to N^k (creating a new matrix $\bar{N}^k$) and P^k is modified accordingly.

Motion is continued in this manner until $P^k \nabla \varphi^k = 0$. Then the gradient corresponding to some negative component (if one exists) of

$$[N^k(N^k)^T]^{-1} N^k \nabla \varphi^k$$

is dropped from N^k, yielding the projection matrix P^k. The algorithm terminates when

$$P^k \nabla \varphi^k = 0 \tag{7.45}$$

and

$$[N^k(N^k)^T]^{-1} N^k \nabla \varphi^k \geq 0. \tag{7.46}$$

These conditions are equivalent to the first-order necessary conditions for x^k to be an optimum. By Theorem 19 these are also sufficient that x^k solve the convex programming problem (in this case φ is required to be a convex function of x).

The vector given by (7.46) consists of the dual variables for the binding constraints (all other dual variables are equal to zero, thus ensuring the complementary slackness condition). Thus (7.45) means that the objective function gradient is equal to a non-negative linear combination of the gradients of the binding constraints.

A proof of convergence of this method and a more detailed explanation are given in [98].

The following example demonstrates how the gradient projection method can be utilized in combination with an unconstrained technique.

Example.

$$\text{minimize } -2x_1 - x_2$$

subject to

$$l_1 = \quad\quad - x_2 + 1 \geq 0,$$
$$l_2 = -4x_1 - 6x_2 + 7 \geq 0,$$
$$l_3 = -10x_1 - 12x_2 + 15 \geq 0,$$
$$g_1 = -x_1 - \frac{x_2^2}{4} + 1 \geq 0,$$
$$x_1 \geq 0, \quad x_2 \geq 0.$$

7.8 Gradient Projection Method 151

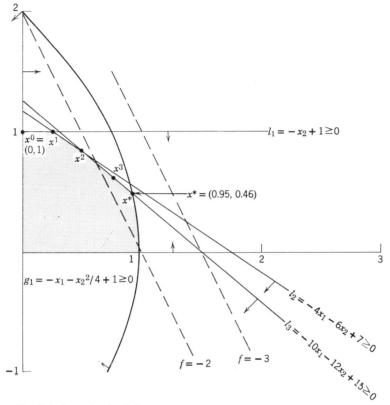

Figure 11 Solution of mixed linearly and nonlinearly constrained problem given in example. The shaded area is the feasible region.

The optimum is at $x^* \doteq (0.95, 0.46)$, as seen in Figure 11. The starting point is assumed to be $x^0 = (0, 1)$. Thus

$$L = -2x_1 - x_2 - r \ln(-x_1 - x_2^2/4 + 1)$$

if we use a logarithmic penalty function to incorporate g_1. We let the initial r value be $r_1 = 1$.

The initial projection matrix takes the form

$$P = \left[I - \begin{pmatrix} 1 & 0 \\ 0 & -1 \end{pmatrix} \begin{pmatrix} 1 & 0 \\ 0 & 1 \end{pmatrix}^{-1} \begin{pmatrix} 1 & 0 \\ 0 & -1 \end{pmatrix} \right] = \begin{bmatrix} 0 & 0 \\ 0 & 0 \end{bmatrix}$$

since the linear constraint l_1 and the non-negativity requirement on x_1 are constraining. The vector of dual variables is

$$\begin{pmatrix} 1 & 0 \\ 0 & -1 \end{pmatrix} \left[\begin{pmatrix} -2 \\ -1 \end{pmatrix} - \frac{4}{3} \begin{pmatrix} -1 \\ -\frac{1}{2} \end{pmatrix} \right] = (-\tfrac{2}{3}, \tfrac{1}{3})^T,$$

which means that we can leave the non-negativity constraint. Removing its gradient from the projection matrix leaves

$$P^0 = \left[I - \begin{pmatrix} 0 \\ -1 \end{pmatrix}(1)(0, -1) \right] = \begin{bmatrix} 1 & 0 \\ 0 & 0 \end{bmatrix}.$$

Projecting the negative gradient of L at $(0, 1)$ gives our first direction, $s^0 = (\frac{2}{3}, 0)^T$.

Now we choose λ^0 so that $L(x^0 - \lambda \nabla L^0, r)$ is minimized, or a linear boundary is pierced, whichever happens first.

$$\min_\lambda \left[-2\lambda(\tfrac{2}{3}) - \ln\left(\frac{-2\lambda}{3} + \frac{3}{4}\right) \right]$$

can be found by differentiating with respect to λ and setting the result equal to zero. Thus $\lambda = \frac{3}{8}$. By chance, the constraint l_2 is encountered at $x^0 - \lambda^0 s^0 = x^1 = (\frac{1}{4}, 1)$.

Also quite by chance, the gradient of $L(x^1, r_1)$ is $(0, 0)$, and we have therefore reached an unconstrained minimum of L for $r_1 = 1$. We now reduce r, and let $r_2 = \frac{1}{4}$. Then $\nabla L(x^1, r_2) = (-\frac{3}{2}, -\frac{3}{4})^T$. We add the gradient of l_2 to P^1, which yields dual variables whose values are given by

$$\frac{1}{16}\begin{pmatrix} 52 & -6 \\ -6 & 1 \end{pmatrix}\begin{pmatrix} 0 & -1 \\ -4 & -6 \end{pmatrix}\begin{pmatrix} -\tfrac{3}{2} \\ -\tfrac{3}{4} \end{pmatrix} = \begin{pmatrix} -\tfrac{3}{2} \\ \tfrac{3}{8} \end{pmatrix}.$$

Thus we can drop the requirement that l_1 be satisfied and project along the l_2 constraint. Our projection matrix is now

$$P^1 = \left[I - \begin{pmatrix} -4 \\ -6 \end{pmatrix}\left(\frac{1}{52}\right)(-4, -6) \right] = \frac{1}{13}\begin{pmatrix} 9 & -6 \\ -6 & 4 \end{pmatrix}.$$

The direction vector is now

$$s_2 = -P^2 \nabla L(x^1, r_2) = (\tfrac{9}{13}, -\tfrac{6}{13})^T.$$

For $\lambda = \frac{13}{36}$, which gives the point where l_3 is encountered, $\partial L/\partial \lambda < 0$, so that the λ for which L is minimized along s_2 is greater than $\lambda = \frac{13}{36}$. We take the smaller λ value and thus $\lambda^1 = \frac{13}{36}$, and $x^2 = (\frac{1}{4}, 1) + \frac{13}{36}(\frac{9}{13}, -\frac{6}{13}) = (\frac{1}{2}, \frac{5}{6})$. We add the gradient of l_3 to the projection matrix, which yields dual variables whose values are equal to

$$\frac{1}{36}\begin{pmatrix} 61 & -28 \\ -28 & 13 \end{pmatrix}\begin{pmatrix} -4 & -6 \\ -10 & -12 \end{pmatrix}\begin{pmatrix} -\tfrac{58}{47} \\ -\tfrac{32}{47} \end{pmatrix} = (-\tfrac{94}{141}, \tfrac{55}{141})^T.$$

Thus l^2 can be eliminated from the projection matrix, which now becomes

$$P^2 = \left[I - \begin{pmatrix} -10 \\ -12 \end{pmatrix}\left(\frac{1}{244}\right)(-10, -12) \right] = \frac{1}{61}\begin{pmatrix} 36 & -30 \\ -30 & 25 \end{pmatrix}.$$

Projecting the negative gradient $-\nabla L(x^2, r_2)$ yields a direction

$$s_2 \doteq (0.393, -0.327).$$

7.8 Gradient Projection Method

No new constraints are encountered in minimizing along this vector, and the optimum λ^2 is ~0.616, yielding $x^3 = (0.742, 0.632)$.

The gradient $\nabla L(x^3, r_2) = (-0.419, -0.500)$ and its projection yields $(-0.001, +0.001)$, which is approximately zero. This point suffices as the constrained minimum of L subject to the linear restrictions.

Reducing r again and minimizing would not, as seen in the geometry of Figure 11, result in a change of constraining linear hyperplanes, but would result in a point closer to the limiting nonlinear constraint. The extrapolation techniques can then be applied and the solution approximated.

The gradient projection method can be modified to take into account the structure of the constraint gradients just as simplicial methods do. Some of the resulting simplicities are as follows.

If at any time during the gradient projection algorithm an inequality of the form $x_j \geq c$ or $x_j \leq c$ is required to be satisfied, the gradient must be added to the N matrix and P modified accordingly. Because of the nature of the gradient, either $(0, \ldots, 0, 1, 0, \ldots, 0)$ or $(0, \ldots, 0, -1, 0, \ldots, 0)$, P takes a special form and several simplifications, computational and structural, are possible. It is intuitively obvious that if P projects every vector into a subspace $x_j = 0$, the jth row (and by symmetry the jth column) of P must be zero. What is not as obvious, but equally as true, is that P is independent of the jth components of all the gradients that make up N. Suppose that the ith row of N is the gradient of all zeros except for a one or minus one in column j. Let $\tilde{N}$ be the same as N with the ith row and jth column deleted. Then it can be shown that $\tilde{P} = [I_{n-1} - \tilde{N}^T(\tilde{N}\tilde{N}^T)^{-1}\tilde{N}]$ is exactly the same as P if the jth row and column of P (which are zeros) are taken out and the dimension of P reduced to $(n-1)$.

To show this we prove the above statement when $x_1 = c_1$ is the relevant constraint. We assume that its gradient is the first one in the N matrix. For any other variable and gradient position the appropriate row and column transformations yield the proper result.

Suppose $N^T = \begin{pmatrix} 1 & a \\ 0 & A \end{pmatrix}$, where $A = \tilde{N}^T$. Then

$$P = I - \begin{pmatrix} 1 & a \\ 0 & A \end{pmatrix}\left[\begin{pmatrix} 1 & 0 \\ a^T & A^T \end{pmatrix}\begin{pmatrix} 1 & a \\ 0 & A \end{pmatrix}\right]^{-1}\begin{pmatrix} 1 & 0 \\ a^T & A^T \end{pmatrix}$$

$$= I - \begin{pmatrix} 1 & a \\ 0 & A \end{pmatrix}\begin{bmatrix} 1 & -a(A^TA)^{-1} A^T \\ 0 & (A^TA)^{-1} A^T \end{bmatrix} \quad (7.47)$$

$$= \begin{bmatrix} 0 & 0 \\ 0 & I - A(A^TA)^{-1}A^T \end{bmatrix}.$$

154 Other Unconstrained Minimization Techniques

To compute the dual variables (7.46), use is made of (7.47).

The following example illustrates the simplicities of computation resulting from the above observations.

Example. It is desired to compute the projection matrix where the relevant constraints are $x_1 = c_1$, $x_3 = c_2$, $2x_1 + x_2 - x_4 = c_3$.

By the regular method,

$$N = \begin{bmatrix} 1 & 0 & 0 & 0 \\ 0 & 0 & 1 & 0 \\ 2 & 1 & 0 & -1 \end{bmatrix},$$

$$(NN^T)^{-1} = \begin{pmatrix} 1 & 0 & 2 \\ 0 & 1 & 0 \\ 2 & 0 & 6 \end{pmatrix}^{-1} = \begin{pmatrix} 3 & 0 & -1 \\ 0 & 1 & 0 \\ -1 & 0 & 0.5 \end{pmatrix},$$

$$(NN^T)^{-1}N = \begin{bmatrix} 1 & -1 & 0 & 1 \\ 0 & 0 & 1 & 0 \\ 0 & 0.5 & 0 & -0.5 \end{bmatrix}, \qquad (7.48)$$

$$N^T(NN^T)^{-1}N = \begin{bmatrix} 1 & 0 & 0 & 0 \\ 0 & 0.5 & 0 & -0.5 \\ 0 & 0 & 1 & 0 \\ 0 & -0.5 & 0 & 0.5 \end{bmatrix}, \text{ and}$$

$$P = \begin{bmatrix} 0 & 0 & 0 & 0 \\ 0 & 0.5 & 0 & 0.5 \\ 0 & 0 & 0 & 0 \\ 0 & 0.5 & 0 & 0.5 \end{bmatrix}. \qquad (4.79)$$

Using the simplified formulas,

$$A^T = (1, -1), \; (A^T A)^{-1} = 0.5,$$

$$(A^T A)^{-1} A^T = (0.5, -0.5), \qquad (7.50)$$

$$A(A^T A)^{-1} A^T = \begin{bmatrix} 0.5 & -0.5 \\ -0.5 & 0.5 \end{bmatrix}, \text{ and}$$

$$I - A(A^T A)^{-1} A^T = \begin{bmatrix} 0.5 & 0.5 \\ 0.5 & 0.5 \end{bmatrix}. \qquad (7.51)$$

7.8 Gradient Projection Method

Equation (7.51) is just the reduced version of (7.49). Note that $a = (2, 0)^T$.
To obtain (7.48) from (7.50) calculate $-a(A^T A)^{-1} A^T = -\begin{pmatrix} 2 \\ 0 \end{pmatrix}(0.5, -0.5) =$
$-\begin{pmatrix} 1 & -1 \\ 0 & 0 \end{pmatrix}$ and use (7.47) to "expand" as

$$\begin{bmatrix} 1 & -1 & 0 & 1 \\ 0 & 0 & 1 & 0 \\ 0 & 0.5* & 0 & -0.5* \end{bmatrix}$$

where the starred numbers are components of (7.50).

In problems with a large number of variables, many of which will be zero at the solution, such a reduction in dimension should effect a great reduction in computation time and storage space. Also, effects of round-off errors are reduced.

8

Computational Aspects of Unconstrained Minimization Algorithms

8.1 INTRODUCTION—SUMMARY OF COMPUTATIONAL ALGORITHM

A vital test and justification of any body of theory of how to solve problems is the feasibility of computational implementation and practical application. In this chapter the computational questions implicit in the theoretical development of unconstrained minimization algorithms are discussed in detail.

The algorithm described will be applicable to any mixed interior-exterior point unconstrained algorithm as defined in Section 4.3. For definiteness we assume that the logarithmic penalty function is to be applied to the constraints where interior feasibility is required, and the quadratic loss function as the exterior point penalty term. The problem to be solved is

$$\text{minimize } f(x) \tag{M}$$

subject to

$$g_i(x) \geq 0, \quad i = 1, \ldots, m, \tag{8.1}$$

$$g_i(x) \geq 0, \quad i = m+1, \ldots, q. \tag{8.2}$$

(Equality constraints may appear among the last $q - m$ constraints.) The unconstrained function has the form

$$W(x, r_k) = f(x) - r_k \sum_{i=1}^{m} \ln g_i(x) + \sum_{i=m+1}^{q} \frac{\{\min [0, g_i(x)]\}^2}{r_k}. \tag{8.3}$$

Most of the computational results are general and apply to any unconstrained function. The sense in which the W function is the "best" mixed unconstrained function is pointed out in later development.

8.2 Minimizing an Unconstrained Function

The computational algorithm is summarized as follows.

1. Find a point x^0 in $R^0 = \{x \mid g_i(x) > 0, i = 1, \ldots, m\}$. If such a point is not readily available the optimization method itself can be used to find it (Section 8.5).

2. Determine r_1, the initial value of r. For the W function we have assumed that the same r value weights all the constraints. As already discussed in Section 7.1, there are various criteria for choosing different weights for each constraint. The discussion when just one r value is used is contained in Section 8.5.

3. Determine the unconstrained minimum of W for the current value of r_k. This step constitutes the bulk of the work required by unconstrained algorithms. Section 8.2 is devoted to ways of minimizing unconstrained functions, and Section 8.3 discusses efficiencies that result when the programming problems have special structure that can be exploited. In Section 8.5 the problem of minimizing a function along a vector is considered. This is a one-dimensional subproblem of the problem of minimizing a function of several variables and can be considered separately.

4. Estimate the solution of Problem M using the extrapolation theory of Chapter 5. The formulas used to do this are contained in Section 8.4.

5. Terminate computations if the estimated solution is "acceptable." A natural criterion for convex programming problems is available via the theory of duality of Section 6.2. Discussion of convergence criteria is given in Section 8.5.

6. Select r_{k+1} (Section 8.5).

7. Estimate the next unconstrained minimum of the W function using the extrapolation formulas of Section 8.4. Continue from Step 3.

8.2 MINIMIZING AN UNCONSTRAINED FUNCTION

Background

Numerous proposals for minimizing an unconstrained function have been made, particularly since Cauchy's "steepest descent" [22] technique was introduced in 1847. Occasional reviews of the proposals have appeared [80, 103, 55], and experience on test problems is published. Unfortunately these reviews are rarely critical. The result is that variations of the old techniques are reused and the same difficulties are often perpetuated. The published experience tends to give the impression that these methods invariably work. Sometimes a failure is published [46], but it is often quickly forgotten.

158 Computational Aspects of Unconstrained Minimization Algorithms

In the first part of this section an attempt is made to give a brief critique of some of these methods. In the second part rational bases for developing general methods for minimizing an unconstrained function are given, and three new methods presented.

It should be emphasized that the problem of minimizing an unconstrained function is not an extraneous difficulty introduced by unconstrained minimization techniques. An unconstrained minimization problem is nothing more than a programming problem with no effective side conditions. Therefore no general method for programming can bypass the essence of the problem of unconstrained minimization. Efforts to develop efficient techniques for unconstrained minimization are thus quite pertinent to the development of mathematical programming methodology.

Without Derivatives

Consider the problem of finding the unconstrained minimum of a continuous function $f(x)$. The derivatives may not be continuous or they may not be explicitly available. This sort of problem can arise when the function values are given by a simulation or possibly a computer-stored table. In many cases, where the function may be differentiable but very involved, coding and round-off error may justify the use of a minimization method requiring only the values of the function.

Methods that do not use derivatives can be divided into at least two classes, search techniques and methods of conjugate directions. The first type is discussed in [103, 109, 80]. A review of the second type is contained in [55].

Many of the search methods lack convergence proofs, are subject to premature convergence, and require manual (as opposed to automatic) parameter adjustment. The published literature for these methods deals mainly with problems having few variables. An interesting example is a small nonconvex test problem, on which the P function [(3.10, Section 3.1] was first tested. This was successfully solved [58, pages 25-26] by a search technique.

Because of their heavy use of univariate probes, search techniques can effectively solve certain problems that give the more conventional gradient techniques appreciable difficulty. The relative efficiency of search techniques generally increases with the "separability" of the functions involved. For example, the problem minimize $[x_1^4 + 10{,}000 x_2^4]$ can be solved in two steps by most search techniques. The isocontours of this function are very elongated ellipses, which make movement along any ray not parallel to the major axis extremely difficult.

Based on the material in [55] there is reason to believe that some newer methods based on moving in "conjugate directions" will apply very successfully to minimizing unconstrained functions of the W type. Powell's method

8.2 Minimizing an Unconstrained Function

[95] is guaranteed to converge to the minimum of a positive definite quadratic form in n^2 iterations. He reports success for problems in 20 variables. A modification of this has been proposed by Zangwill [116].

Most of the problems solved by the authors have had at least first partial derivatives that were analytically computable and reasonably tractable. Because of this the methods above have not been explored in any depth. Development of methods in this area is important, since it should permit a user the option of furnishing function values only, rather than requiring the calculation of first derivatives, in utilizing a nonlinear programming algorithm. An unconstrained minimization technique requiring only function values is presented in the second half of this section.

With First Derivatives

Three methods for minimizing an unconstrained function that use first partial derivatives are presented here. One of the oldest and best known is the method of "steepest descent," first proposed by Cauchy [22]. Iterations are made according to the following equation:

$$x^{i+1} = x^i - \lambda^i \nabla f(x^i), \tag{8.4}$$

where λ^i is the smallest non-negative value of λ that locally minimizes f along $-\nabla f(x^i)$ starting from x^i. It follows from (8.4) that successive gradients in the procedure are orthogonal; that is, for all i,

$$\nabla^T f(x^{i+1}) \nabla f(x^i) = 0.$$

Curry [32] showed that any limit point $\bar{x}$ of the sequence $\{x^i\}$ is a stationary point; that is, $\nabla f(\bar{x}) = 0$. Under stronger conditions, Goldstein [66] has proved this also.

Unfortunately, "steepest descent" has several disadvantages that often make it a poor method for minimizing an unconstrained function. These are as follows.

1. It is not generally a finite procedure for minimizing a positive definite quadratic form.
2. Each iteration is calculated independently of the others; that is, no information is stored and used that might accelerate convergence.
3. The rate of convergence depends strongly on the graph of the function. If the ratio of the maximum to minimum eigenvalues of the matrix of second partial derivatives of the function at any solution point is large, steepest descent generates short zigzagging moves in a neighborhood of such a point. Thousands of such moves may be required to give a satisfactory approximation of a limit point [49, page 41].

Various attempts to accelerate convergence in these cases have been made, often without rigorous basis. All gradient methods for solving nonlinear programming problems must eventually deal with this convergence problem.

The rationale for the next two methods is the minimization in n or fewer steps of a positive definite quadratic form. These methods are based on generating vectors in "conjugate directions."

Suppose that f is of the form

$$f(x) = b^T x + \frac{x^T G x}{2}, \qquad (8.5)$$

where G is a positive definite matrix. Let $s^0, \ldots, s^{n-1}$ be nonzero vectors and the λ^i be scalars with the property that

$$f(x^{k+1}) = f\left[x^0 + \sum_{i=0}^{k-1} \lambda^i s^i + \lambda^k s^k \right]$$

is the minimum of f along s^k, starting from the point $x^k = x^0 + \sum_{i=0}^{k-1} \lambda^i s^i$, for $k = 0, \ldots, n-1$. The s^i are called conjugate, with respect to the matrix G, if

$$(s^i)^T G s^j = 0, \qquad i \neq j. \qquad (8.6)$$

It follows that x^n is the unconstrained minimum of f in the entire space. To prove this, note first that the positive definiteness of G, the fact that the s^i are nonzero, and (8.6) imply that the s^i are independent. It follows that

$$\nabla f(x^n) = \nabla f\left[x^k + \sum_{i=k}^{n-1} \lambda^i s^i \right], \qquad k = 0, \ldots, n-1,$$

$$= b + G[x^k + \lambda^k s^k] + \sum_{i=k+1}^{n-1} \lambda^i G s^i, \qquad k = 0, \ldots, n-1,$$

$$= \nabla f[x^{k+1}] + \sum_{i=k+1}^{n-1} \lambda^i G s^i, \qquad k = 0, \ldots, n-1.$$

Multiplying both sides by $(s^k)^T$ yields

$$(s^k)^T \nabla f(x^n) = (s^k)^T \nabla f(x^{k+1}) + (s^k)^T \sum_{i=k+1}^{n-1} \lambda^i G s^i = 0$$

by the orthogonality of consecutive moves generated by choice of the λ^i and conjugacy of the s^i, for $k = 0, \ldots, n-1$. Since the s^k are independent, it follows that $\nabla f(x^n) = 0$, a sufficient condition that x^n be the minimum.

The methods to be presented are ways of generating successive conjugate directions.

8.2 Minimizing an Unconstrained Function

Fletcher and Reeves [57] proposed the following. Let x^0 be the starting point, let $s^0 = -\nabla f(x^0)$, let x^{i+1} be a point yielding the minimum value of $f(x)$ on the line from x^i in direction s^i, and let

$$\beta^i = |\nabla f(x^{i+1})|^2 / |\nabla f(x^i)|^2.$$

Then the new direction s^{i+1} is given by

$$s^{i+1} = -\nabla f(x^{i+1}) + \beta^i s^i.$$

The directions s^i, $i = 0, \ldots, n-1$ are conjugate if $f(x)$ is of the form (8.5).

Thus for quadratic functions of the above form the algorithm will locate the minimum in n steps regardless of the starting point. For general functions Fletcher and Reeves suggest restarting the algorithm every $(n+1)$ steps.

The third method, called the "variable metric" method, is a modification by Fletcher and Powell [56] of a technique proposed by W. C. Davidon [35]. In addition to n-step convergence for a positive definite quadratic, the inverse of the matrix of second partial derivatives at the optimum is generated. (This means storage of the order of $n^2/2$ locations.) The algorithm is as follows. Let x^0 be the initial point and H^0 an arbitrary positive definite matrix. Let

$$s^i = -H^i \nabla f(x^i),$$

$$\sigma^i = \lambda^i s^i,$$

$$x^{i+1} = x^i + \sigma^i,$$

where λ^i is chosen to minimize f along s^i, starting at x^i. If H^i is positive definite, λ^i must be greater than zero unless x^i is a minimum of $f(x)$. Let

$$y^i = \nabla f(x^{i+1}) - \nabla f(x^i).$$

The new approximation of the inverse of the matrix of second partial derivatives of f is given by

$$H^{i+1} = H^i + \frac{(\sigma^i)(\sigma^i)^T}{(\sigma^i)^T y^i} - \frac{H^i(y^i)(y^i)^T H^i}{(y^i)^T H^i y^i}.$$

It is shown in [56] that H^{i+1} is positive definite if H^i is positive definite, ensuring that f is decreased on each move. The vectors $\sigma^0, \ldots, \sigma^{n-1}$ are conjugate with respect to the matrix G if f is given by (8.5). A method similar to this, containing several additional properties, is given in the second part of this section.

Using Second Partial Derivatives

The most successful method used to date by the authors for minimizing an unconstrained convex function is the generalized Newton method, which requires second partial derivatives. For notational convenience let $F(x^i)$ represent $\nabla^2 f(x^i)$. It is assumed that $F(x^i)$ is nonsingular. The iterations are given by

$$x^{i+1} = x^i - \lambda^i F(x^i)^{-1} \nabla f(x^i), \qquad (8.7)$$

where $\lambda^i > 0$ is chosen to minimize f along $-F(x^i)^{-1} \nabla f(x^i)$, starting from x^i. With $\lambda^i \equiv 1$, this is called Newton's method. The motivation for this is seen by first expanding the function in a two-term Taylor series and solving for the vector, which yields the minimum of the quadratic approximation. A fuller discussion may be found in [31].

If f is a positive definite quadratic form (8.7) with $\lambda^i = 1$ will yield the minimum in one step. When f is a convex function (not necessarily quadratic) (8.7) will guarantee a decrease in f with every iteration if x^i is not a minimum.

The disadvantages of the generalized Newton, or "second-order," method are the following.

1. An inverse of $F(x^i)$ may not always exist.
2. In a nonconvex problem the function may not decrease when the current point x^i is not near the minimum.
3. For some problems with continuous second partial derivatives it may be tedious and troublesome to compute the analytic forms accurately.
4. Storage of the order of $n^2/2$ locations is required.

Nonetheless, for problems in which second partial derivatives are readily available, the method has proved far superior to any other tried. In Section 8.3 the computation of (8.7) for the logarithmic quadratic penalty function in many special cases of mathematical programming is shown to be of much less than the expected order of $n^3/3$ operations. This follows from the special form of the matrix of second partial derivatives of this function.

A revised Newton method overcoming the first two of the above criticisms is discussed in the second half of this section.

Comparison of the Methods

To do justice to these methods would require much more space, analysis, and experimentation. We shall give a small example that was solved using each of the four methods just discussed to illustrate their convergence properties.

Example. The function to be minimized is

$$P(x, r) = f(x) + r \sum_{i=1}^{10} \frac{1}{g_i} + r \sum_{j=1}^{5} \frac{1}{x_j}, \qquad r = 1, x \in R^0$$

where f and $\{g_i\}$ are the functions given in Section 7.3. In the region where $R^0 = \{x \mid g_i(x) > 0, i = 1, \ldots, 10, x_j > 0 \text{ all } j\}$, $P(x, 1)$ is a *strictly convex*

8.2 Minimizing an Unconstrained Function

function. Tables 6, 7, 8, and 9 summarize the "work" required to minimize $P(x, 1)$ by each of the four techniques just described.

Table 6 Gradient Descent on Example

Iteration	x_1	x_2	x_3	x_4	x_5	$|\nabla_x P|$	$P(x, 1)$
0	10^{-4}	10^{-4}	10^{-4}	10^{-4}	1.1		
1	0.0798	0.0798	0.0798	0.0798	1.1000	6×10^2	88.919014
10	0.1194	0.0694	0.1932	0.2001	1.0040	1×10^2	44.533934
20	0.1732	0.0982	0.2772	0.3285	0.8691	8×10^1	21.944428
30	0.1835	0.1344	0.2887	0.4086	0.7786	5×10^1	14.884308
50	0.1814	0.1726	0.2898	0.4913	0.6670	3×10^1	10.084040
100	0.1776	0.2260	0.2889	0.5537	0.5157	1×10^1	7.329493
125	0.1762	0.2395	0.2898	0.5605	0.4775	8×10^0	7.058207
151	0.1763	0.2477	0.2894	0.5642	0.4543	4×10^0	6.960957

Table 7 Fletcher-Reeves Method on Example

Iteration	x_1	x_2	x_3	x_4	x_5	$|\nabla_x P|$	$P(x, 1)$
0	10^{-4}	10^{-4}	10^{-4}	10^{-4}	1.1		
5	0.1921	0.0865	0.2794	0.3517	0.9154	2×10^2	24.626489
10	0.1747	0.2321	0.2921	0.6065	0.5376	4×10^1	7.812791
15	0.1781	0.2435	0.2883	0.5460	0.4629	1×10^1	7.036639
20	0.1750	0.2555	0.2879	0.5735	0.4350	2×10^0	6.926209
30	0.1764	0.2573	0.2889	0.5700	0.4278	3×10^{-1}	6.919394
40	0.1764	0.2575	0.2890	0.5699	0.4272	4×10^{-2}	6.919354

Table 8 Variable Metric Method on Example

Iteration	x_1	x_2	x_3	x_4	x_5	$|\nabla_x P|$	$P(x, 1)$
0	10^{-4}	10^{-4}	10^{-4}	10^{-4}	1.1		
5	0.0651	0.0833	0.0933	0.0775	0.9584	4×10^2	75.359926
10	0.0646	0.0643	0.0998	0.0907	0.9128	4×10^2	73.072791
15	0.1601	0.0937	0.2521	0.2162	0.8253	1×10^2	27.639421
20	0.1857	0.2229	0.2864	0.5966	0.5013	7×10^1	7.772631
25	0.1736	0.2578	0.2862	0.5713	0.4342	5×10^0	6.933100
28	0.1764	0.2575	0.2889	0.5699	0.4271	8×10^{-3}	6.919352

The method of steepest descent failed to converge after 151 iterations when the iterations were manually terminated. The method of Fletcher and Reeves took 40 iterations to converge, the variable metric method, 28; and the "second order" method, 8. The effort per iteration is approximately the same

Table 9 Generalized Newton or "Second-Order" Method on Example

Iteration	x_1	x_2	x_3	x_4	x_5	$|\nabla_x P|$	$P(x, 1)$
0	10^{-4}	10^{-4}	10^{-4}	10^{-4}	1.1		
1	8.2×10^{-4}	8.2×10^{-4}	8.2×10^{-4}	8.2×10^{-4}	1.4204	3×10^6	5.00×10^3
2	0.0501	0.0501	0.0501	0.0501	0.9296	1×10^3	114.854
3	0.1749	0.1962	0.2596	0.2381	0.6438	4×10^2	23.263122
4	0.1979	0.2532	0.3165	0.4213	0.4637	2×10^2	11.293004
5	0.1786	0.2885	0.2957	0.5896	0.3888	7×10^1	7.512202
6	0.1766	0.2577	0.2882	0.5632	0.4380	9×10^0	6.947183
7	0.1764	0.2580	0.2890	0.5702	0.4263	5×10^{-1}	6.919448
8	0.1764	0.2575	0.2889	0.5699	0.4271	7×10^{-3}	6.919353

for each method, so that the required computer time is in about the same ratio as the number of iterations.

The tremendous power of the second order method for problems of this kind (as exemplified by this example), that is, for convex programming problems, has been seen by the authors in many computer solved examples. In this example the $n^3/3$ operations (multiplications and additions) required to compute the direction vector for each iteration of the second-order method did not significantly increase the time required to solve the problem, since n was so small. For "large" problems this cubic law can become prohibitive. Some techniques for reducing the computational effort are discussed in Section 8.3.

Even though the Fletcher-Reeves and variable metric methods require on the order of n and $n^2/2$ operations, respectively, to compute the direction vectors, both require approximately n times as many moves to converge. (This is to be expected, since the second-order method minimizes a positive definite form in one step requiring $n^3/3$ operations, and the others require n iterations of n and $n^2/2$ operations, respectively.)

It is worth examining the Fletcher-Reeves method a little more closely, since the above remarks seem to imply that the number of operations required to minimize a positive definite quadratic form varies as n^2 rather than n^3. However, the fact is that the amount of work required to find the minimum of a quadratic along a vector is of the order of n^2 operations.

The area of investigation on ways to minimize an unconstrained function has only been touched on. Much good work has been done in recent years in this field. Modifications of the last three methods discussed will probably accelerate the process for problems with special characteristics or structure. For the second-order method, Section 8.3 contains an attempt to utilize the special structure of the W function to simplify the second-order computations.

In the last part of this section we present three methods for minimizing unconstrained functions, one for each order of differentiability assumed.

Three New Methods for Minimizing an Unconstrained Function Using a Second-Order Point of View

In this part of the section we discuss three new methods of minimizing an unconstrained function. We discuss, first, a modified Newton method that uses second partial derivatives; second, a modified variable metric method that uses first partials and can take advantage of second partial information if available; and, third, a method that employs only function values.

As already mentioned, two traditional theoretical arguments against the generalized Newton method are, first, if $F(x^i)$ is not a positive definite matrix, a move in the direction given by (8.7) may result in an increase rather than a decrease in $f(x)$, yielding $\lambda^i = 0$, and terminate the process at x^i. The second theoretical objection is that $F(x^i)$ may not have an inverse, even if $f(x)$ is convex.

The modified second-order method to be presented takes into account these two relevant objections to the generalized Newton method. The direction vector s^i is generated according to two rules. In both cases $x^{i+1} = x^i + \lambda^i s^i$, where λ^i is chosen to be the smallest value of $\lambda \geq 0$ for which $x^i + \lambda^i s^i$ is a local minimum of $f(x^i + \lambda s^i)$. The rules are as follows.

1. If F^i has a negative eigenvalue, let s^i be a vector where

$$(s^i)^T F^i s^i < 0 \text{ and } (s^i)^T \nabla f^i \leq 0. \tag{8.8}$$

2. If F^i has all eigenvalues greater than or equal to zero, choose s such that either

$$F^i s = 0, \qquad s^T \nabla f^i < 0 \tag{8.9}$$

or

$$F^i s = -\nabla f^i \tag{8.10}$$

(in which case, if $\nabla f^i \neq 0$, it can be proved that $s^T F^i s = -s^T \nabla f^i > 0$). Both (8.9) and (8.10) cannot hold. Let s^i satisfy (8.9) or (8.10).

The only case in which Rules 1 and 2 fail to generate a nonzero direction s^i is when F^i is a positive semidefinite matrix and $\nabla f^i = 0$; that is, a point that satisfies the first- and second-order necessary conditions that x^i be a local constrained minimum of $f(x)$.

The rationale for Rule 1 is that if the second partial derivative matrix has a negative eigenvalue there are certain directions along which the function $f(x)$ decreases and along which the *rate* of decrease also decreases. That is, for a vector s satisfying (8.8), $df(x^i + \lambda s)/d\lambda = s^T \nabla f^i \leq 0$, and

$$d^2 f(x^i + \lambda s)/d\lambda^2 = s^T F^i s < 0$$

at $\lambda = 0$. Thus, unless f varies along the ray s^i emanating from x^i so that eventually $d^2 f(x^i + \lambda s)/d\lambda^2 \geq 0$, the function value will tend to $-\infty$. This rule provides a guide for locating a region where F^i is positive semidefinite. For the minimization of a nonconvex function a few iterations along vectors satisfying (8.8) can result in significant progress in a minimization algorithm.

The same remarks apply to (8.9) of Rule 2. If $F(x^i + \lambda s)s = 0$ holds for λ indefinitely large, $f(x^i + \lambda s)$ will tend to $-\infty$. Finally, if F^i is positive definite, (8.10) yields the Newton vector. When (8.10) is used, in general,

$$s = -(F^i)^\# \nabla f^i,$$

where $(F^i)^\#$ is a generalized inverse of F^i satisfying $(F^i)(F^i)^\# F^i = F^i$. We note that any s satisfying (8.10) has the desirable property that if $f(x)$ is a

8.2 Minimizing an Unconstrained Function

positive semidefinite quadratic form having a finite unconstrained minimum, then $x^i + s$ is such an unconstrained minimum.

Rules 1 and 2, although quite useful computationally, are not guaranteed to generate a sequence of points having limit points satisfying the first- and second-order necessary conditions for a minimum. A class of direction vectors for which any existing limit point has this property is established in [86]. One particular example of this type of direction vector is

$$s^i = -A \nabla f^i + \delta^i e^i,$$

where A is a positive definite matrix, e^i is an eigenvector of F^i with minimum eigenvalue, and δ^i is a scalar, where

$$\delta_i = \begin{cases} -1 & \text{if } F^i \text{ has a negative eigenvalue and } (\nabla f^i)^T e^i \geq 0, \\ 0 & \text{if } F^i \text{ is positive semidefinite,} \\ 1 & \text{if } F^i \text{ has a negative eigenvalue and } (\nabla f^i)^T e^i < 0. \end{cases}$$

From the above observations, the suggested algorithm is given as follows.

Modified Newton Method

1. Reduce the matrix F^i to the form

$$F^i = L^i D^i (L^i)^T,$$

where L^i is a nonsingular lower triangular matrix and D^i is a diagonal matrix.

2. If D^i has all positive diagonal elements, solve for $s^i = -(F^i)^{-1} \nabla f^i$.

3. If D^i has some diagonal elements that are negative, solve $(L^i)^T t = a^i$, where a^i is a column vector with jth component $= 0$ if the jth diagonal element of $D^i > 0$ and with jth component $= 1$ if the jth diagonal element of $D^i \leq 0$. Let $s^i = t$ if $t^T \nabla f^i \leq 0$, and $s^i = -t$ otherwise. Observe that s^i satisfies (8.8).

4. If D^i has all nonnegative diagonal elements, and at least one is zero, the vector s is generated according to (8.9) or (8.10).

We give two examples of the application of this algorithm. The first example to be solved is

minimize $100(x_2 - x_1^2)^2 + (1 - x_1)^2 + 90(x_4 - x_3^2)^2 + (1 - x_3)^2$

$+ 10.1[(x_2 - 1)^2 + (x_4 - 1)^2] + 19.8(x_2 - 1)(x_4 - 1).$

The starting point of the process is given as $(-3, -1, -3, -1)$.

The problem was submitted by C. F. Wood of Westinghouse Research Laboratories as a test problem for an evaluation of nonlinear programming algorithms sponsored by A. R. Colville of IBM. It was designed to force most algorithms to be trapped at a nonoptimal stationary point. This indeed did happen in many cases. The usual descent and Newton procedures failed to solve the problem; the sequence of points converged roughly to $(-1.07, 1.16, -0.86, 0.76)$, where F was approximately $+7.89$. Somewhere in this region the gradient of F tends to zero. The second partial derivative matrix, however, has a negative eigenvalue and the usual Newton procedure fails to take advantage of this.

The modifications discussed above were applied to the problem with the following results. As is seen in Table 10, 24 iterations were required to solve the problem. Of these, 22 used the usual Newton direction, and just two, from Points 3 to 4 and Points 6 to 7, used Equation 8.8. It was these two moves, however, that enabled the algorithm to get around the difficult corner.

A simpler example illustrates the power of the algorithm more dramatically. Consider the problem

$$\text{minimize } (x_1^4 - 3)^2 + x_2^4.$$

As a starting point $x^0 = (10^{-3}, 10^2)$ was used. The "steepest descent" algorithm immediately led to a region near the point $(0, 0)$ and after several thousand iterations failed to move significantly away from that point though clearly not converging to $(0, 0)$.

The modified Newton method just discussed used the fact that

$$H^0 \doteq \begin{pmatrix} 1 & 0 \\ 0 & 1 \end{pmatrix} \begin{pmatrix} -72 \times 10^{-6} & 0 \\ 0 & 12 \times 10^2 \end{pmatrix} \begin{pmatrix} 1 & 0 \\ 0 & 1 \end{pmatrix}$$

has a negative diagonal element and moved orthogonal to the x_2 direction, as prescribed by the choice of direction vector as indicated above. The point x^1 was $\doteq (\sqrt[4]{3}, 10)$, where the second partial derivative matrix was positive definite. After 8 moves the remainder of them being Newton, the point $(1.316074, 1.53507 \times 10^{-6})$ with an objective function value of 8.88×10^{-16} was given as the solution. The actual solution is $(\sqrt[4]{3}, 0)$.

The second method presented here is a modification of the variable metric method. This method is similar in that it requires only the first partial derivatives of the function at any point. The main difference is that it requires only *one* term to be added at each iteration to approximate the inverse of the second partial derivative matrix. First, we prove convergence of the modified n-step procedure for the quadratic function (8.5). Second, we show how, if assumptions are made involving second partial derivatives, the steps required to minimize can usually be reduced.

Table 10

Iteration	x_1	x_2	x_3	x_4	Type of Move	F
0	−3.0000	−1.0000	−3.0000	−1.0000		19,192.00
1	−2.6873	6.4007	−2.5528	6.0793	N†	1,275.324
2	−1.7697	2.0543	−1.7893	2.1960	N	273.2618
3	−1.3084	1.8623	−1.2356	1.5115	(8.8)	31.49227
4	−0.7916	0.5673	−1.5336	2.2529	N	17.87550
5	−0.5954	0.3228	−1.2813	1.5875		8.354000
6	−0.5467	0.3099	−1.2586	1.5969	(8.8)	7.772888
7	−0.4150	0.1674	−1.3106	1.7275	N	7.705842
8	−0.2601	0.0523	−1.3546	1.8426	N	7.591963
9	−0.0375	−0.0223	−1.3808	1.9165	N	7.297546
10	0.2184	0.0148	−1.3678	1.8801	N	6.791865
11	0.5073	0.2163	−1.2946	1.6806	N	5.998664
12	0.8138	0.6192	−1.1388	1.2863	N	4.933558
13	1.0905	1.1620	−0.8985	0.7737	N	3.844972
14	1.2844	1.6461	−0.6113	0.3363	N	2.978971
15	1.3753	1.8902	−0.3168	0.0670	N	2.325651
16	1.4157	2.0069	−0.0369	−0.0318	N	1.769503
17	1.3906	1.9336	0.2398	0.0235	N	1.217641
18	1.3228	1.7480	0.5172	0.2351	N	6.635734×10^{-1}
19	1.1901	1.4026	0.770	0.5723	N	2.217372×10^{-1}
20	1.0498	1.1012	0.9641	0.9235	N	1.645591×10^{-2}
21	1.0076	1.0137	0.9921	0.9839	N	5.582219×10^{-4}
22	1.0004	1.0008	0.9997	0.9994	N	1.084665×10^{-6}
23	1.0000003	1.0000005	0.99999976	0.99999945	N	1.123790×10^{-12}
Theoretical solution	1.0	1.0	1.0	1.0		0.0

† N = Newton method.

Revised Variable Metric Method

Suppose that x^0 is the initial point of the process. Let H^0 be an arbitrary symmetric positive semidefinite matrix. This may be thought of as an initial approximation of the inverse of the matrix of second partial derivatives. Let s^i denote a nonzero direction vector at the ith iteration, where s^i is linearly independent of the previous direction vectors $s^0, s^1, \ldots, s^{i-1}$. One final requirement is that $\nabla^T f^i s^i < 0$, all i. Clearly, unless $\nabla f^i = 0$ at the ith iteration at least one s^i exists with those properties.

Let $\sigma^i = \lambda^i s^i$, where λ^i is chosen to minimize f along s^i starting at x^i. If $\nabla^T f^i s^i < 0$, λ^i must be greater than 0. Let

$$y^i = \nabla f^{i+1} - \nabla f^i,$$
$$x^{i+1} = x^i + \sigma^i. \tag{8.11}$$

The new approximation to the inverse of the matrix of second partial derivatives is then given by

$$H^{i+1} = H^i + (\sigma^i - H^i y^i)[(y^i)^T(\sigma^i - H^i y^i)]^{-1}((\sigma^i)^T - (y^i)^T H^i). \tag{8.12}$$

We assume that $(y^i)^T(\sigma^i - H^i y^i) \neq 0$. If it is equal to zero, no new approximation of H^i is made, but new directions are chosen. We assume that $(y^i)^T(\sigma^i - H^i y^i)$ will not equal zero indefinitely.

After n changes of H, the direction $s^n = -H^n \nabla f^n$ is used, and, as will be proved in the following theorem, x^{n+1} is the desired unconstrained minimum of a positive definite quadratic form.

The rationale for modifying H according to (8.12) is simply that if f is a quadratic form, it follows that

$$\sigma^i = G^{-1} y^i. \tag{8.13}$$

We want to modify our current approximation of G^{-1} so that it at least satisfies (8.13). A very simple modification is obtained by adding a symmetric matrix of rank 1. Thus we require a vector z and a scalar δ such that

$$[H^i + z\delta z^T]y^i = \sigma^i, \tag{8.14}$$

yielding

$$z[\delta z^T y^i] = \sigma^i - H^i y^i.$$

Thus z must be proportional to $\sigma^i - H^i y$. (See [15] for a general discussion of similar developments.)

We would also like the previous equations of type (8.14) to hold for the new approximation H^{i+1}; that is, we want $H^{i+1} y^j = \sigma^j$ for $0 \leq j \leq i$. Then at the nth iteration we have

$$H^n y^i = \sigma^i, \quad i = 0, 1, \ldots, n-1.$$

8.2 Minimizing an Unconstrained Function

Since the scalar δ can be adjusted to any value, depending on our choice for z, we let $z = \sigma^i - H^i y^i$, yielding $\delta = [(y^i)^T(\sigma^i - H^i y^i)]^{-1}$. Then, with H^{i+1} designating the matrix in brackets in (8.14), we obtain (8.12). Because the σ^i are independent, we shall show that $H^n = G^{-1}$.

Theorem 41 [Convergence in $n + 1$ Steps for a Positive Definite Quadratic Form]. Assume that $f(x)$ is a positive definite quadratic form, where $\nabla^2 f(x) = G$. Assume that $(\sigma^0, \sigma^1, \ldots, \sigma^{n-1})$ are n linearly independent vectors for which (8.12) and (8.13) hold and for which

$$\gamma^i = (y^i)^T(\sigma^i - H^i y^i) \neq 0, \quad i = 0, \ldots, n-1.$$

Then $H^n = G^{-1}$.

Proof. The proof is inductive. We first show that $H^1 y^0 = \sigma^0$.

Let $z^i = \sigma^i - H^i y^i$. Using (8.12),

$$H^1 y^0 = [H^0 + z^0(\gamma^0)^{-1}(z^0)^T]y^0$$
$$= H^0 y^0 + z^0(\gamma^0)^{-1}(\gamma^0) = H^0 y^0 + \sigma^0 - H^0 y^0 = \sigma^0.$$

Assume $H^i y^j = \sigma^j$, $0 \leq j \leq i-1$. Then we show that $H^{i+1} y^j = \sigma^j$, $0 \leq j \leq i$.

$$H^{i+1} y^i = H^i y^i + z^i (\gamma^i)^{-1}(\gamma^i)$$
$$= H^i y^i + \sigma^i - H^i y^i = \sigma^i.$$

When $j < i$,

$$H^{i+1} y^j = H^i y^j + z^i(\gamma^i)^{-1}(\sigma^i - H^i y^i)^T y^j$$
$$= \sigma^j + z^i(\gamma^i)^{-1}[(\sigma^i)^T y^j - (y^i)^T \sigma^j]$$

(using the induction hypothesis twice)

$$= \sigma^j + z^i(\gamma^i)^{-1}[(\sigma^i)^T G \sigma^j - (\sigma^i)^T G \sigma^j]$$

[using (8.13)]

$$= \sigma^j.$$

Then $H^n y^j = \sigma^j$, $0 \leq j \leq n-1$. Since $G^{-1} y^j = \sigma^j$, $0 \leq j \leq n-1$ by (8.13), and since the σ^i are assumed linearly independent, we can conclude that $H^n = G^{-1}$. Q.E.D.

Example. This example is worked out in detail to illustrate the flexibility in choice of s^i and H^0 available. Consider the quadratic programming problem in E^2.

$$\text{minimize } (1, 1)x + \tfrac{1}{2} x^T \begin{pmatrix} 2 & 1 \\ 1 & 1 \end{pmatrix} x.$$

Let $x^0 = (0\ 0)^T$, $H^0 = \begin{pmatrix} 0 & 0 \\ 0 & 0 \end{pmatrix}$. Then $\nabla f^0 = (1, 1)^T$. We let $s^0 = (-1, 0)^T$

Then $\lambda^0 = \frac{1}{2}$, $\sigma^0 = (-\frac{1}{2}, 0)^T$,

$$x^1 = (0, 0)^T + (-\frac{1}{2}, 0)^T = (-\frac{1}{2}, 0)^T, \quad \nabla f^1 = (0, \frac{1}{2})^T,$$

$$y^0 = (0, \frac{1}{2})^T - (1, 1)^T = (-1, \frac{1}{2})^T,$$

$$\sigma^0 - H^0 y^0 = (-\frac{1}{2}, 0)^T, \quad (y^0)^T(\sigma^0 - H^0 y^0) = \frac{1}{2},$$

$$H^1 = \begin{pmatrix} -\frac{1}{2} \\ 0 \end{pmatrix} 2(-\frac{1}{2}, 0) = \begin{pmatrix} \frac{1}{2} & 0 \\ 0 & 0 \end{pmatrix}.$$

Let $s^1 = (0, -1)$. Then $\lambda^1 = \frac{1}{2}$, $\sigma^1 = (0, -\frac{1}{2})^T$,

$$x^2 = (-\tfrac{1}{2}, 0)^T + (0, -\tfrac{1}{2})^T = (-\tfrac{1}{2}, -\tfrac{1}{2})^T, \quad \nabla f^2 = (-\tfrac{1}{2}, 0)^T,$$

$$y^1 = (-\tfrac{1}{2}, 0)^T - (0, \tfrac{1}{2})^T = (-\tfrac{1}{2}, -\tfrac{1}{2})^T,$$

$$\sigma^1 - H^1 y^1 = \begin{pmatrix} 0 \\ -\frac{1}{2} \end{pmatrix} - \begin{pmatrix} \frac{1}{2} & 0 \\ 0 & 0 \end{pmatrix} \begin{pmatrix} -\frac{1}{2} \\ -\frac{1}{2} \end{pmatrix} = \begin{pmatrix} \frac{1}{4} \\ -\frac{1}{2} \end{pmatrix},$$

$(y^1)^T(\sigma^1 - H^1 y^1) = \frac{1}{8}$,

$$H^2 = \begin{pmatrix} \frac{1}{2} & 0 \\ 0 & 0 \end{pmatrix} + \begin{pmatrix} \frac{1}{4} \\ -\frac{1}{2} \end{pmatrix} 8(\tfrac{1}{4}, -\tfrac{1}{2}) = \begin{pmatrix} 1 & -1 \\ -1 & 2 \end{pmatrix},$$

which is the correct inverse of $\begin{pmatrix} 2 & 1 \\ 1 & 1 \end{pmatrix}$. The next move yields the unconstrained minimum,

$$x^3 = \begin{pmatrix} -\frac{1}{2} \\ -\frac{1}{2} \end{pmatrix} + \begin{pmatrix} 1 & -1 \\ -1 & 2 \end{pmatrix} \begin{pmatrix} \frac{1}{2} \\ 0 \end{pmatrix} = \begin{pmatrix} 0 \\ -1 \end{pmatrix}.$$

Another significant modification of the variable metric method will be introduced that utilizes information about a submatrix of the matrix of second partial derivatives. When this information is available, say when an $n - r$ square submatrix of second partial derivatives is known, only $(r + 1)$ steps of the variable metric method are needed to minimize a positive definite quadratic form (under appropriate assumptions).

For simplicity we assume that $\partial^2 f/\partial x_i \partial x_j$, $i, j = 1, \ldots, n - r$ are known. We denote the square submatrix of second partial derivatives with respect to these first $n - r$ variables as $\tilde{G}$. Then

$$G = \nabla^2 f = \begin{bmatrix} \tilde{G} & a \\ a^T & b \end{bmatrix}. \tag{8.15}$$

8.2 Minimizing an Unconstrained Function

Since G is assumed to be positive definite, and hence its inverse exists, the inverse of the bordered matrix G is

$$G^{-1} = \begin{pmatrix} \tilde{G}^{-1} & 0 \\ 0 & 0 \end{pmatrix} + \begin{pmatrix} -\tilde{G}^{-1}a \\ I \end{pmatrix}(-a^T\tilde{G}^{-1}a + b)^{-1}(-a^T\tilde{G}^{-1}, I). \quad (8.16)$$

The second term in (8.16) is an outer product matrix of rank r, which we propose to determine in r steps. That, coupled with the fact that $\tilde{G}^{-1}$ is assumed known, will enable us to calculate G^{-1}.

The only difference in the present procedure is to let

$$H^0 = \begin{pmatrix} \tilde{G}^{-1} & 0 \\ 0 & 0 \end{pmatrix}. \quad (8.17)$$

The change from H^k to H^{k+1} is exactly as stated before. The important difference is that H^k is positive *semi*definite but *never* positive *definite* until the full inverse is obtained.

Because of the foregoing we are ready to prove the following theorem.

Theorem 42 [Use of Second Derivative Information]. If $f(x) = b^T x + x^T G x/2$ is to be minimized, G is a positive definite quadratic form, and $\tilde{G}$ as given by (8.15) is known, H^0 is given by (8.17), and H^{i+1} is given by (8.12); if $\sigma^0, \ldots, \sigma^{r-1}$ are r linearly independent vectors satisfying (8.13), for which

$$\gamma^i = (y^i)^T(\sigma^i - H^i y^i) \neq 0, i = 1, \ldots, r - 1,$$

and if $A^T Y$ has rank r, (where $A^T = (-a^T \tilde{G}^{-1}, I), Y \equiv (y^0, \ldots, y^{r-1}))$, then

$$H^r = G^{-1}$$

and the indicated revised procedure yields a minimum in $(r + 1)$ steps.

Proof. Using the same method of proof as in Theorem 41 we know that

$$H^r y^j = \sigma^j, \quad (8.18)$$

where for $0 \leq j \leq r - 1$ the σ^j are linearly independent vectors. We also know that

$$H^r = \begin{pmatrix} \tilde{G}^{-1} & 0 \\ 0 & 0 \end{pmatrix} + X, \quad (8.19)$$

where X is an $n \times n$ symmetric matrix. Now $Y = (y^0, \ldots, y^{r-1})$ and $\sigma = (\sigma^0, \ldots, \sigma^{r-1})$. From (8.16) and (8.13) we know that

$$\left[\begin{pmatrix} \tilde{G}^{-1} & 0 \\ 0 & 0 \end{pmatrix} + ADA^T\right] Y = \sigma, \quad (8.20)$$

where A is $n \times r$ and has rank r, and D is $r \times r$, has rank r, and is symmetric (and by a brief argument D must be positive definite).

We shall prove that $X = ADA^T$, and it then follows that $H^r = G^{-1}$. Using (8.18), (8.19), and (8.20),

$$XY = \sigma - \begin{pmatrix} \tilde{G}^{-1} & 0 \\ 0 & 0 \end{pmatrix} Y = ADA^T Y. \tag{8.21}$$

Multiplying both sides of (8.21) by Y^T yields

$$Y^T X Y = Y^T ADA^T Y. \tag{8.22}$$

From (8.20) we know that Y has rank r, and from the assumption on $A^T Y$ we know that X must be a positive semidefinite matrix of rank r. Hence X must be factorable as ZZ^T, where Z is $n \times r$. There are an infinite number of such factorizations, but we show that ZZ^T is invariant and hence that $X = ADA^T$.

We prove the following lemma.

Lemma 16. Let Z be any $n \times r$ matrix satisfying

$$ZZ^T Y = (ADA^T)Y = V, \tag{8.23}$$

where V has rank r and Y is an $n \times r$ matrix of rank r. Then the product ZZ^T is invariant.

Proof. Multiplying both sides of (8.23) by Y^T, we have

$$(Y^T Z)(Z^T Y) = Y^T ADA^T Y = U. \tag{8.24}$$

Because of our assumptions the matrix on the right-hand side of (8.24) is an $r \times r$ positive definite matrix. Thus any $Y^T Z$ is of the form

$$Y^T Z = CO, \tag{8.25}$$

where $COO^T C^T = Y^T ADA^T Y$ and $OO^T = I$, C and O being square matrices. Because Z satisfies (8.25) it must be of the form

$$Z = Y(Y^T Y)^{-1} CO + [I - Y(Y^T Y)^{-1} Y^T]v, \tag{8.26}$$

where v is arbitrary. Substituting (8.26) into (8.23) and reducing yields

$$[I - Y(Y^T Y)^{-1} Y^T]v = [I - Y(Y^T Y)^{-1} Y^T]V(O^T C^T)^{-1}. \tag{8.27}$$

8.2 Minimizing an Unconstrained Function

Substituting the left side of (8.27) into (8.26) and multiplying Z by Z^T yields

$$ZZ^T = Y(Y^TY)^{-1}V^TP^Y + P^YV(U)^{-1}V^TP^Y$$
$$+ Y(Y^TY)^{-1}(U)(Y^TY)^{-1}Y^T + P^YV(Y^TY)^{-1}Y^T,$$

where $P^Y = [I - Y(Y^TY)^{-1}Y^T]$. This is the required invariant form.

Q.E.D.

The lemma allows us to conclude that $H^r = G^{-1}$.

The obvious advantage of this algorithm is that it allows the variable metric method to be used on large problems with a structure that makes it convenient or possible to compute second partial derivatives for some of the variables and not for others. The fact that a second-order move can be approximated in fewer than n steps may be a tremendous advantage in solving these large problems.

An example of the above mixed second-order–first-order move follows.

Example. The problem is the same as the previous example,

$$\text{minimize } (1, 1)x + \tfrac{1}{2}x^T \begin{pmatrix} 2 & 1 \\ 1 & 1 \end{pmatrix} x .$$

We assume that g_{11} is known to be 2, and let $H^0 = \begin{pmatrix} \tfrac{1}{2} & 0 \\ 0 & 0 \end{pmatrix}$. Let $x^0 = (1, -2)^T$. Then $\nabla f^0 = (1, 0)^T$. Let $s^0 = (-1, -1)^T$. Then $\lambda^0 = \tfrac{1}{5}$, $\sigma^0 = (-\tfrac{1}{5}, -\tfrac{1}{5})^T$,

$$x^1 = (1, -2)^T + (-\tfrac{1}{5}, -\tfrac{1}{5})^T = (\tfrac{4}{5}, -\tfrac{11}{5})^T, \qquad \nabla f^1 = (\tfrac{2}{5}, -\tfrac{2}{5})^T,$$

$$y^0 = (\tfrac{2}{5}, -\tfrac{2}{5})^T - (1, 0)^T = (-\tfrac{3}{5}, -\tfrac{2}{5})^T,$$

$$\sigma^0 - H^0 y^0 = \begin{pmatrix} -\tfrac{1}{5} \\ -\tfrac{1}{5} \end{pmatrix} - \begin{pmatrix} \tfrac{1}{2} & 0 \\ 0 & 0 \end{pmatrix} \begin{pmatrix} -\tfrac{3}{5} \\ -\tfrac{2}{5} \end{pmatrix} = \begin{pmatrix} \tfrac{1}{10} \\ -\tfrac{2}{10} \end{pmatrix}, \qquad \gamma^0 = \tfrac{1}{50},$$

$$A^1 = \begin{pmatrix} \tfrac{1}{2} & 0 \\ 0 & 0 \end{pmatrix} + \begin{pmatrix} \tfrac{1}{10} \\ -\tfrac{2}{10} \end{pmatrix} 50(\tfrac{1}{10}, -\tfrac{2}{10}) = \begin{pmatrix} 1 & -1 \\ -1 & 2 \end{pmatrix},$$

which is the required inverse after one iteration.

Method Using Only Function Values

The third new method for minimizing an unconstrained function involves only the evaluation of the function values, not the first or second partial derivatives.

As with the modified variable metric method, this method is exact for a positive definite quadratic form. It is based on the idea of approximating the second partial derivative matrix G (and its gradient), using $n(n + 1)/2$ iterations. The final iteration is a Newton second-order move.

Assume that $f(x)$ is a quadratic form. Let G denote $\nabla^2 f(x)$. Then

$$f(x^{k+1}) = f(x^k) + \nabla^T f^k \sigma^k + \frac{\sigma_k^T G \sigma_k}{2}, \tag{8.28}$$

where

$$\sigma^k = x^{k+1} - x^k. \tag{8.29}$$

Assume that an optimum step from x^k to x^{k+1} has been made; then

$$\nabla^T f^{k+1} \sigma_k = 0. \tag{8.30}$$

Thus

$$f(x^k) = f(x^{k+1}) - \nabla^T f^{k+1} \sigma_k + \tfrac{1}{2} \sigma_k^T G \sigma_k$$

$$= f(x^{k+1}) + \frac{\sigma_k^T G \sigma_k}{2}. \tag{8.31}$$

Then, using (8.28) and (8.31),

$$\nabla^T f^k \sigma_k = -\sigma_k^T G \sigma_k. \tag{8.32}$$

Rewriting (8.31) yields

$$f(x^{k+1}) - f(x^k) = -\frac{\sigma_k^T G \sigma_k}{2}. \tag{8.33}$$

Using (8.33) we can obtain the diagonal elements of G by letting our first n direction vectors be the vectors $(1, 0, \ldots, 0), \ldots, (0, \ldots, 0, 1)$. Then the obvious formulas are used to get the off-diagonal elements of G.

More specifically the algorithm is as follows. Let x^0 be the starting point. For $i = 1, \ldots, n$ let s^{i-1} be a vector with one in the ith position and zeros elsewhere. Then $x^{i+1} = x^i - \lambda^i s^i$, where λ^i (which may be positive or negative) is chosen to minimize f on the vector s^{i-1} emanating from x^{i-1}. The diagonal elements of G are obtained as

$$g_{i+1, i+1} = \frac{-2(f^{i+1} - f^i)}{(\lambda^i)^2}, \quad i = 0, \ldots, n - 1. \tag{8.34}$$

To obtain the off-diagonal elements of G we proceed as follows. Let $x^{k_0} = x^n$ (just generated by the n probes to calculate the diagonal elements). For $i = 1, \ldots, n$ and $j = i + 1, \ldots, n$, let

$$s^{i,j} = (e^i + e^j),$$

where e^i is a vector with all zeros except a one in position i. Let $\lambda^{i,j}$ be the

8.2 Minimizing an Unconstrained Function

scalar that minimizes f along $s^{i,j}$ starting from x^k. Then

$$g_{i,j} = \frac{2(-f^{k+1} + f^k) - (\lambda^{i,j})^2 g_{ii} - (\lambda^{i,j})^2 g_{jj}}{2(\lambda^{ij})^2}. \tag{8.35}$$

It is possible that the λ's chosen above can be zero if the quadratic form is minimized in the direction chosen. In this case a small arbitrary probe can be made and the formulas above still apply.

To get the proper gradient observe the following.

After a unit probe in the ith direction ($s^{i-1} = e^i$), $\partial f(x^i)/\partial x_i = 0$. Now

$$\frac{\partial f(x^n)}{\partial x_i} = \frac{\partial f(x^i)}{\partial x_i} + \sum_{j=1}^{n} g_{i,j}(x_j^n - x_j^i)$$

assuming that f is quadratic. But

$$x_j^n = x_j^i, \quad j = 1, \ldots, i \quad \text{and} \quad x_j^i = x_j^0, \quad j = i+1, \ldots, n.$$

Hence

$$\frac{\partial f(x^n)}{\partial x_i} = \sum_{j=i+1}^{n} g_{i,j}(x_j^n - x_j^0). \tag{8.36}$$

Thus the gradient at x^n is computable when all elements of G are obtained. For any other point y,

$$\nabla f(y) = \nabla f(x^n) + G(y - x^n). \tag{8.37}$$

After G has been computed, (8.37) can be used to obtain $\nabla f(y)$ and a direction $-G^{-1} \nabla f(y)$ generated. The point $y - G^{-1} \nabla f(y)$ will be the unconstrained minimum of a positive definite quadratic form.

The solution of the following small problem illustrates the use of the algorithm.

Example. Consider the problem

$$\text{minimize } (1, 1)x + \tfrac{1}{2} x^T \begin{pmatrix} 2 & 1 \\ 1 & 1 \end{pmatrix} x.$$

Let $x^0 = (0, 0)$. Then $f^0 = 0$. Let $s^0 = (1, 0)$. Then $\lambda^0 = -\tfrac{1}{2}$, $\sigma^0 = (-\tfrac{1}{2}, 0)^T$, $x^1 = (-\tfrac{1}{2}, 0)$, and $f^1 = -\tfrac{1}{4}$. Using (8.34),

$$g_{1,1} = -2(-\tfrac{1}{4} - 0)/(\tfrac{1}{2})^2 = 2.$$

Let $s^1 = (0, 1)$. Then $\lambda^1 = -\tfrac{1}{2}$, $\sigma^1 = (0, -\tfrac{1}{2})$, $x^2 = (-\tfrac{1}{2}, -\tfrac{1}{2})$, and $f^2 = -\tfrac{3}{8}$. Then $g_{2,2} = -2(-\tfrac{3}{8} + \tfrac{1}{4})/(-\tfrac{1}{2})^2 = 1$. Let $s^2 = (1, 1)$. Then $\lambda^3 = \tfrac{1}{10}$, $x^3 = (-\tfrac{2}{5}, -\tfrac{2}{5})$, and $f^3 = -\tfrac{2}{5}$. Using (8.35),

$$g_{1,2} = \frac{2(\tfrac{2}{5} - \tfrac{3}{8}) - (\tfrac{1}{10})^2(2 + 1)}{2(1/10)^2}$$

$$= 1.$$

All elements of G have been properly identified. To get ∇f^2 we use (8.36),

$$\frac{\partial f(x^2)}{\partial x_1} = g_{1,2}(x_2^2 - x_2^0) = (1)(-\tfrac{1}{2} + 0) = -\tfrac{1}{2},$$

$$\frac{\partial f(x^2)}{\partial x_2} = 0 \text{ (applying (8.36) in a logical sense)}.$$

Using (8.37),

$$\nabla f^3 = \begin{pmatrix} -\tfrac{1}{2} \\ 0 \end{pmatrix} + \begin{pmatrix} 2 & 1 \\ 1 & 1 \end{pmatrix} \left[\begin{pmatrix} -\tfrac{2}{5} \\ -\tfrac{2}{5} \end{pmatrix} + \begin{pmatrix} \tfrac{1}{2} \\ \tfrac{1}{2} \end{pmatrix} \right]$$

$$= \begin{pmatrix} -\tfrac{1}{5} \\ \tfrac{1}{5} \end{pmatrix},$$

which is correct. Then

$$x^3 - G^{-1} \nabla f^3 = \begin{pmatrix} -\tfrac{2}{5} \\ -\tfrac{2}{5} \end{pmatrix} - \begin{pmatrix} 1 & -1 \\ -1 & 2 \end{pmatrix} \begin{pmatrix} -\tfrac{1}{5} \\ \tfrac{1}{5} \end{pmatrix}$$

$$= \begin{pmatrix} 0 \\ -1 \end{pmatrix}$$

is the required unconstrained minimum.

8.3 MINIMIZING THE W FUNCTION FOR CONVEX PROGRAMMING PROBLEMS WITH SPECIAL STRUCTURE

We now explore the application of some of the second-order methods discussed in the previous sections to problems having special structure.

As mentioned in the preceding section, the generalized Newton method, in the authors' experience thus far, has been computationally established as the most effective method for minimizing an unconstrained convex function. The difficulty with it is that each computation of (8.7) takes the order of $n^3/6$ operations. Also, storage of $n^2/2$ locations is required. Thus the number of variables n in the problem is the controlling factor in determining the speed and size of problems that can be handled by the method. This is in contrast to linear programming methods, where the number of nontrivial constraints (non-negativity or "trivial" requirements assumed) is generally the controlling factor in determining the number of iterations (which is proportional to the time) required, and the size of problems that can be handled.

In this chapter special cases of convex programming—linear programming, separable programming, and "factorable" programming—are examined in detail. Most "real-world" convex problems seen by the authors have fallen

8.2 Minimizing an Unconstrained Function

into these categories. Knowledge that a function is convex generally means that its matrix of second partial derivatives is factorable and that it can be handled in factored form more easily. The precise definition of factorable is given later.

In many cases considerable reduction in effort, in computing a second-order iteration, can be had by using the rank annihilation method of inverting a matrix [114, 72].† This method can also be used to take advantage of the sparseness of the constraint gradients.

Rank Annihilation Method of Inverting Matrices

The rank annihilation method of inverting matrices can be summarized by the following equation:

$$[A + u\sigma v^T]^{-1} = A^{-1} - A^{-1}u[\sigma^{-1} + v^T A^{-1} u]^{-1} v^T A^{-1}. \quad (8.38)$$

This can be easily validated by multiplying the right side of (8.38) by the matrix in brackets on the left. Here A is an $n \times n$ matrix, u is an $n \times p$ matrix, σ is a $p \times p$ matrix, and v^T is a $p \times n$ matrix. We assume that the required inverses exist. Under the circumstances in which (8.38) will be used, all the necessary requirements will be met.

Equation 8.38 gives a powerful way of systematically getting an inverse by successively modifying existing ones. The necessary requirement is that the matrix to be inverted be written as the sum of an invertible matrix A and an outer product or "proportional" matrix. It turns out that this is the form of the matrix of second partial derivatives of the W function, since this is given by

$$\nabla^2 W(x, r) = \nabla^2 f(x) - \sum_{i=1}^{m} \frac{r}{g_i(x)} \nabla^2 g_i(x)$$

$$+ [\nabla g_1(x), \ldots, \nabla g_m][\text{diag } (r/g_i^2(x))] \begin{bmatrix} \nabla^T g_1(x) \\ \cdot \\ \cdot \\ \cdot \\ \nabla^T g_m(x) \end{bmatrix}$$

$$+ \sum_{i=m+1}^{q} \delta_i \frac{2g_i(x)}{r} \nabla^2 g_i(x)$$

$$+ [\nabla g_{m+1}(x), \ldots, \nabla g_q(x)] \text{ diag}\left(\frac{2\delta_i}{r}\right) \begin{bmatrix} \nabla^T g_{m+1}(x) \\ \cdot \\ \cdot \\ \cdot \\ \nabla^T g_q(x) \end{bmatrix}, \quad (8.39)$$

---

† We are indebted to S. B. Noble for first bringing these references to our attention.

where $\delta_i = 0$ if $g_i(x) \geq 0$, and $\delta_i = 1$ otherwise. The third and fifth terms on the right of (8.39) are in proportional form.

If any interior point penalty term were of the form $I_i(g_i) = g_i^{-\epsilon}$, for any $\epsilon > 0$ [such as the P function (3.10), where $\epsilon = 1$], it would require more computations than the logarithmic term to calculate the second partial derivative matrix and the vector of first partial derivatives. This is the reason that we have reverted to using the logarithmic penalty function.

If any exterior point penalty term were of the form $O_i(g_i) = [\min(0, g_i)]^{1+\epsilon}$, for any positive $\epsilon \neq 1$, the first partial derivative vector and the second partial derivative matrix would be more difficult to compute. This is the reason for using the quadratic loss function in lieu of any other penalty term of the form indicated.

Notice also that a linear constraint adds a *constant* proportional matrix of the form $\nabla g_i (2\delta_i/r) \nabla^T g_i$ to $\nabla^2 W$ when absorbed in the form $[\min(0, g_i)]^2/r$, whereas it is $\nabla g_i [r/g_i^2(x)] \nabla^T g_i$ if absorbed in the form $-r \ln g_i$. This latter matrix is not constant and depends on the current value of $g_i(x)$. This fact can be used to reduce the computations of $\nabla^2 W$ and it implies that linear constraints should be handled by the quadratic loss function. This is seen in the discussion of linear programming that follows.

We now examine how the special cases of mathematical programming problems mentioned above can utilize (8.38) and (8.39). Our emphasis is on calculating the information to make a Newton iteration (8.7), or a modified Newton iteration [(8.7) and (8.10)].

Linear Programming

The linear programming problem in E^n can be written

$$\text{minimize } c^T x$$

subject to

$$Ax - b \geq 0, \tag{8.40}$$

$$x_j \geq 0, \quad j = 1, \ldots, n, \tag{8.41}$$

$$Ex - d = 0, \tag{8.42}$$

where A is $m \times n$ and E is $p \times n$. We apply the quadratic loss function to absorb the constraints. For the equality constraints (8.42) recall from the discussion in Section 4.3 that the corresponding penalty term takes the form $|Ex - d|^2/r$. The matrix of second partial derivatives of the W function for the linear programming problem is

$$\nabla^2 W^k = \left[\text{diag}\left(\frac{2\delta_j^k}{r}\right) \right] + A^T \left[\text{diag}\left(\frac{2\delta_i^k}{r}\right) \right] A + E^T \left[\text{diag}\left(\frac{2}{r}\right) \right] E, \tag{8.43}$$

where $\delta_j^k = 0$ if $x_j \geq 0$, $\delta_j^k = 1$ otherwise; $\delta_i^k = 0$ if $a_i x - b_i \geq 0$, $\delta_i^k = 1$ otherwise; and a_i is the ith row of A.

8.3 Minimizing the W Function for Structured Convex Problems

The operations required to compute the inverse of (8.43) depend on the number of constraints of the form (8.41) and (8.42) that are not satisfied at the current point, and the number of equality constraints. Important facts about (8.43) are that (a) advantage can be taken of its constant structure to update the inverse for every second-order move rather than recomputing it afresh each time, and (b) if the inverse exists, sparseness of A and E and the structure of $[\text{diag}(\delta_j/r)]$ can be used to simplify the computations.

If at any iteration k there are at least n independent vectors in (8.43), the second partial derivative matrix can be represented as $\nabla^2 W(x^k, r) = (2/r)[LL^T + MM^T]$, where L is a square matrix of rank n. Applying (8.38) to this yields

$$(\nabla^2 W^k)^{-1} = \frac{r}{2}\left\{(LL^T)^{-1} - (LL^T)^{-1}M[I + M^T(LL^T)^{-1}M]^{-1}M^T(LL^T)^{-1}\right\}. \tag{8.44}$$

Given (8.44), a second-order Newton iteration (8.7) is now possible. If fewer than n independent vectors are available, then (9.9) and (8.10) can be used. For example, if $\nabla^2 W(x^k, r) = (2/r)[NN^T]$, where N^T is $q \times n$, and has rank q when q is less than n, then, using (8.9),

$$s = -\left[I - N(N^T N)^{-1} N^T\right]\left[c + N\tilde{g}\left(\frac{2}{r}\right)\right]$$

$$= -[I - N(N^T N)^{-1} N^T]c, \tag{8.45}$$

where $\tilde{g}$ is a $q \times 1$ vector of constraint values whose corresponding gradients compose N. This is the same type of move given by the projected gradient method. In the rare event that s as given by (8.45) is the zero vector, (8.10) must be used to get the direction vector yielding

$$s = -\frac{r}{2}[N(N^T N)^{-1}(N^T N)^{-1} N^T]\left[c + N\tilde{g}\left(\frac{2}{r}\right)\right].$$

In (8.44), $(LL^T)^{-1} = (L^T)^{-1}L^{-1}$, and L^{-1} can be computed by any method that takes advantage of its structure. For example, if a non-negativity condition $x_j \geq 0$ is not satisfied, one column vector of L is zero except for a one in the jth position. The inverse has the same form in the jth column. The considerable work used to invert sparse matrices in recent years can also be employed to invert L efficiently.

The same remarks apply to (8.45). In the discussion of modifications to Rosen's projected gradient method (Section 7.8), it was shown how $(N^T N)^{-1} N^T$ can be computed more efficiently when N contains column vectors with only one nonzero element. When computing the optimum λ^k for the second-order moves similar efficiencies can be obtained.

After an optimal move the inverse can be updated directly using (8.38). If some constraint at x^{k+1} is violated that was satisfied at x^k, this adds a term $v(2/r)v^T$ (v is $n \times 1$ and is the gradient of that constraint) to the matrix of second partial derivatives. Thus the new inverse is

$$(\nabla^2 W^{k+1})^{-1} = (\nabla^2 W^k)^{-1} - (\nabla^2 W^k)^{-1} v \left(\frac{r}{2} + v^T \nabla^2 W^k v\right)^{-1} v^T (\nabla^2 W^k)^{-1}. \quad (8.46)$$

This updating takes on the order of n^2 operations. Subtracting a vector that corresponds to a constraint violated at iteration k and satisfied at iteration $k + 1$ would require an analogous computation.

Similar remarks hold for changing (8.45). The following example illustrates the solution of a linear programming problem using second-order moves and the quadratic loss function. The computational efficiencies suggested are not shown here, since they are applicable only to much larger problems.

Example.

$$\text{minimize } -3x_1 - 2x_2$$

subject to

$$-x_2 + \tfrac{5}{2} \geq 0,$$
$$-x_1 - x_2 + 3 \geq 0,$$
$$-2x_1 - x_2 + 4 \geq 0.$$

(There are no non-negativity restrictions in this problem.)

ITERATION 1

$$W(x^0, 1) = 0, \quad x^0 = (0, 0), \quad \nabla W^0 = \begin{pmatrix} -3 \\ -2 \end{pmatrix}, \quad \nabla^2 W^0 = \begin{pmatrix} 0 & 0 \\ 0 & 0 \end{pmatrix}.$$

Hence (8.45) yields $s^0 = \begin{pmatrix} 3 \\ 2 \end{pmatrix}$. The λ that minimizes $W(x^0 + \lambda s^0, 1)$ is $\lambda^0 = 107/178 \doteq 0.60112$.

$W(x^1, 1) = -5.409$, $x^1 = (1.803, 1.202)$,

$$\nabla W^1 = -\begin{pmatrix} 3 \\ 2 \end{pmatrix} + 2(-0.005)\begin{pmatrix} -1 \\ -1 \end{pmatrix} + 2(-0.808)\begin{pmatrix} -2 \\ -1 \end{pmatrix}$$

$$= \begin{pmatrix} 0.242 \\ -0.374 \end{pmatrix} \quad \nabla^2 W^1 = 2\left[\begin{pmatrix} -1 \\ -1 \end{pmatrix}(-1, -1) + \begin{pmatrix} -2 \\ -1 \end{pmatrix}(-2, -1)\right]$$

$$= 2\begin{pmatrix} 5 & 3 \\ 3 & 2 \end{pmatrix}.$$

$$(\nabla^2 W^1)^{-1} = \frac{1}{2}\begin{pmatrix} 2 & -3 \\ -3 & 5 \end{pmatrix}, \quad s^1 = \frac{1}{2}\begin{pmatrix} 2 & -3 \\ -3 & 5 \end{pmatrix}\begin{pmatrix} -0.242 \\ 0.374 \end{pmatrix} = \begin{pmatrix} -0.803 \\ 1.298 \end{pmatrix}.$$

$\lambda^1 = 1$, $x^2 = (1.0, 2.5)$, which is the unconstrained minimum of $W(x^0, 1)$.

8.3 Minimizing the W Function for Structured Convex Problems

This suggests the facility in solving linear problems using the quadratic loss exterior function.

Separable Programming

By separable nonlinear programming we mean that class of problems that can be written

$$\text{minimize } f(x) = \sum_{j=1}^{n} f_j(x_j)$$

subject to

$$g_i(x) = \sum_{j=1}^{n} g_{ji}(x_j) - b_i \geq 0, \quad i = 1, \ldots, m, \quad (8.47)$$

$$x_j \geq 0, \quad j = 1, \ldots, n. \quad (8.48)$$

Separability implies that all the off-diagonal terms of the matrix of second partial derivatives of any problem function are zero. This is a generalization of its use in the literature, where "separable" is usually meant to apply to a problem with linear constraints and separable objective function.

For simplicity in presentation we assume that the interior of the set defined by (8.47) and (8.48) is nonempty, and that the logarithmic interior point penalty term is used to treat these constraints. Then

$$\nabla^2 W(x, r) = D_1 + N D_2 N^T, \quad (8.49)$$

where

$$D_1 = \text{diag}\left\{\frac{r}{x_j^2} + \frac{\partial^2 f_j(x_j)}{\partial x_j^2} - \sum_{i=1}^{m} \frac{r}{g_i(x)} \frac{\partial^2 g_{ji}(x)}{\partial x_j^2}\right\}, \quad (8.50)$$

$$N = [\nabla g_1(x), \ldots, \nabla g_m(x)], \quad D_2 = \text{diag}\left[\frac{r}{g_i^2(x)}\right].$$

Using (8.38), the inverse is given by

$$[\nabla^2 W(x, r)]^{-1} = D_1^{-1} - D_1^{-1} N [D_2^{-1} + N^T D_1^{-1} N]^{-1} N^T D_1^{-1}. \quad (8.51)$$

Since D_1 is a diagonal matrix only n divisions are required to obtain D_1^{-1}. {The convexity of f and the concavity of the g_i are always assumed in this discussion, ensuring the existence of D_1^{-1} and $[\nabla^2 W(x, r)]^{-1}$ when the $x_j \geq 0$ constraints are present in the problem.}

Although the inverse can be computed from (8.51), in this form it requires more than $n^2/2 + m^2/2$ locations of storage, the computing of a matrix inversion of order m, and a triple matrix product. Also, in this form, it is not possible to take advantage of the sparseness of N (the matrix of gradients). Instead, using (8.38), the following iterative scheme can be used for inverting $\nabla^2 W(x, r)$.

Let v^i represent an $n \times 1$ vector with components $\{v^i(j)\}$, and let v_c^i be an associated scalar. Let

$$d_1^{-1}(j) = \left\{ \frac{r}{x_j^2} + \frac{\partial^2 f_j(x_j)}{\partial x_j^2} - \sum_{i=1}^{m} \frac{r}{g_i(x)} \frac{\partial^2 g_{ji}(x)}{\partial x_j^2} \right\}^{-1}, \quad j = 1, \ldots, n. \quad (8.52)$$

For $i = 1, \ldots, m$, iterate as follows.

$$t^i(j) = d_1^{-1}(j) \cdot \frac{\partial g_{ji}(x)}{\partial x_j}, \quad j = 1, \ldots, n,$$

$$v^i = t^i - \sum_{k=1}^{i-1} v^k [v_c^k \cdot (v^k)^T \nabla g_i(x)],$$

$$v_c^i = \left\{ \left[\frac{r}{g_i^2(x)} \right]^{-1} + \nabla^T g_i(x) v^i \right\}^{-1}. \quad (8.53)$$

It follows that

$$[\nabla^2 W(x, r)]^{-1} = \text{diag}\, [d_1^{-1}(j)] - \sum_{i=1}^{m} v^i (v_c^i)(v^i)^T.$$

Using this iterative scheme, which is simply the repeated application of (8.38), $n(m+1) + n$ storage locations are required for $[\nabla^2 W(x, r)]^{-1}$. If the vectors v^i, $i = 1, \ldots, m$ are sparse, it is possible to reduce this storage. In the example at the end of this section, this procedure is used.

The number of operations required using (8.53) is of the order of nm^2. With m substantially smaller than n, this method of computing the inverse should effect a substantial reduction in computing time. Further computational reductions are possible in (8.53) if the gradients are sparse.

Unlike the development for linear programming, where the use of the quadratic loss function allowed for a few changes for updating $\nabla^2 W(x, r)$ from point to point, these computations must be completely redone each time, since the second partial derivative matrix depends on the function values at the current point.

Factorable Programming

By factorable nonlinear convex programming is meant the class of problems with nonlinear functions having second partial derivatives *given* in a special factorable form. If the problem is written

$$\text{minimize } f(x)$$

subject to

$$g_i(x) \geq 0, \quad i = 1, \ldots, m,$$

$$x_j \geq 0, \quad j = 1, \ldots, n,$$

8.3 Minimizing the W Function for Structured Convex Problems

then the problem is factorable if

$$\nabla^2 f = F(x) \operatorname{diag}[f_k(x)] F^T(x), \quad f_k(x) \neq 0, \, k = 1, \ldots, L, \tag{8.54}$$

$$\nabla^2 g_i = -G_i(x) \operatorname{diag}[\gamma_{i,k}] G_i^T(x),$$

$$\gamma_{i,k} \neq 0, \, k = 1, \ldots, L_i, \, i = 1, \ldots, m. \tag{8.55}$$

Since f and $\{-g_i\}$ are assumed to be convex, twice continuously differentiable functions, their matrices of second partial derivatives evaluated at any specific point can theoretically always be written this way. This is *not* what is intended here, however. A factorable problem is one where the analytic derivation of the second partial derivative matrix directly yields the form of (8.54) and (8.55).

For the factorable problem (treating all constraints with the logarithmic penalty function),

$$\nabla^2 W(x, r) = D_1 + N D_2 N^T,$$

$$[\nabla^2 W(x, r)]^{-1} = D_1^{-1} - D_1^{-1} N [D_2^{-1} + N^T D_1^{-1} N]^{-1} N^T D_1^{-1},$$

where $D_1 = \operatorname{diag}(r/x_j^2)$, and

$$N = [\nabla g_1(x), \ldots, \nabla g_m(x),$$

$$F(x), G_1(x), \ldots, G_m(x)],$$

and

$$D_2 = \begin{bmatrix} \operatorname{diag}\left[\dfrac{r}{g_i^2(x)}\right] & & \\ & \operatorname{diag}[f_k(x)] & \\ & & \operatorname{diag}(\alpha_i) \end{bmatrix},$$

where

$$\alpha_i = \dfrac{r}{g_i^2(x)} \begin{bmatrix} \gamma_{i,1} & & & \\ & \cdot & & \\ & & \cdot & \\ & & & \gamma_{i,L_i} \end{bmatrix}.$$

This assumes the most complicated situation, namely, that all the constraints and the objective function are nonlinear. The obvious modifications are made when any constraint is linear.

There is, of course, no reason precluding the simultaneous use of the separability techniques discussed previously if some of the functions are separable. The complete procedure for a mixed separable-factorable problem proceeds as follows.

Include in the matrix D_1 all matrices in the expansion $\nabla^2 W(x, r)$ that are diagonal. This matrix is easily inverted. Then iterate on the matrices $\sum_{i=1}^{m} \nabla g_i(x)[r/g_i^2(x)]\nabla^T g_i(x)$ and the proportional matrices resulting from the "factorable" constraints using the scheme of (8.52) and (8.53).

Most real world convex problems seen by the authors are factorable. Several well-known problems in the published literature illustrate this.

One of the best known "quadratic" programming problems is Markowitz's portfolio selection problem [83]. The objective function to be minimized is the variance of a probability distribution and is usually given as $f(x) = x^T C x$, where C is a positive semidefinite variance-covariance matrix. But C is a matrix computed from empirical observations. Let M be an $n \times l$ matrix of observations, where l is the number of time periods. Then m_{jk} is the value of the jth variable at the kth time period. The variance-covariance matrix C is given by

$$C = MM^T/l - M(e_l)e_l^T M^T, \qquad (8.56)$$

where e_l is an $l \times 1$ vector of ones. Since $\nabla^2 f(x) = 2C$, $f(x)$ is a factorable function with $l + 1$ factors. From the point of view of the above discussion, using unconstrained minimization algorithms it is reasonable to work with the matrix of original data. (This avoids, incidentally, the $nl^2/2$ computations required to calculate C.)

Another standard objective function is of the form

$$\text{minimize } f(x) = \sum_{j=1}^{q} w_j \exp\left\{-\sum_{i=1}^{p} \alpha_{ij} x_{ij}\right\}. \qquad (8.57)$$

The $\{x_{ij}\}$ are double subscripted variables ($p \cdot q$ in number), the w_j are strictly positive constants, and the α_{ij} are non-negative constants for all i, j. (See [14, Chapter 2] for an example.) The function (8.57) is convex and factorable. The exact form of the second partial derivative matrix is the sum of q factors, each factor having only p nonzero elements. If the variables are ordered as $x = (x_{11}, \ldots, x_{p1}, \ldots, x_{1q}, \ldots, x_{pq})$, then

$$\nabla^2 f(x) = \sum_{j=1}^{q} w_j \alpha_j^T \exp\left\{-\sum_{i=1}^{p} \alpha_{ij} x_{ij}\right\} \alpha_j,$$

where $\alpha_j = \{0, \ldots, 0, \alpha_{1j}, \ldots, \alpha_{pj}, 0, \ldots, 0\}$.

Other examples of factorable convex nonlinear functions are to be found in [107] and [24].

8.3 Minimizing the W Function for Structured Convex Problems

Example [The Chemical Equilibrium Problem]. To illustrate how the development discussed in the preceding sections can reduce the computational effort required to solve nonlinear programming problems, a specific example—the chemical equilibrium problem—was solved. The form of the problem is as follows.

$$\text{minimize } f(x) = \sum_{k=1}^{q} \left\{ \sum_{j=1}^{n_k} x_{jk} \left[c_{jk} + \ln \left(\frac{x_{jk}}{\sum_{i=1}^{n_k} x_{ik}} \right) \right] \right\}$$

subject to

$$Hx = b,$$

$$x \geq 0.$$

[The vector $x = (x_{1,1}, \ldots, x_{n_1,1}, \ldots, x_{1,q}, \ldots, x_{n_q,q})$.] For a description of the physical meaning of the problem the reader is referred to [109].

The matrix of second partial derivatives of f has the form

$$\nabla^2 f = \text{diag}\left(\frac{1}{x_j}\right) + \sum_{k=1}^{q} I_k \left(\frac{-1}{\sum_{i=1}^{n_k} x_{ik}} \right) I_k^T,$$

where I^k is an $n \times 1$ vector with ones in the positions associated with $(x_{1,k}, \ldots, x_{n_k,k})$ as it appears in the vector x, and zeros elsewhere. The mixed interior–exterior point unconstrained function used to solve this was $P = f + r \sum 1/x_j + \sum (h_i^T x - b_i)^2/r$, whose second partial matrix is

$$\nabla^2 P = \text{diag}\left(\frac{1}{x_j} + \frac{2r}{x_j^3}\right) + \sum_{i=1}^{p} h_i \left(\frac{2}{r}\right) h_i^T + \sum_{k=1}^{q} I_k \left(\frac{-1}{\sum_{i=1}^{n_k} x_{ik}} \right) I_k^T \quad (8.58)$$

where h_i^T is the ith row of H. The matrix $(\nabla^2 P)^{-1}$ was obtained by the method just described. Iteration using formula (8.53) was done $p + q$ times, the order as indicated in (8.58). Since consecutive vectors resulting from (8.53) are sparse, they were stored in packed form.

The data were taken from [36, pages 174 and 175]. For this problem $n = 45$ (11 of the 56 variables in [36] are kept at zero and thus do not enter in the computations), $p = 16$, and $q = 7$. The matrix H has 88 nonzero elements.

Using the regular sequential unconstrained minimization technique (SUMT) computer code [85] each computation of $(\nabla^2 P)^{-1} \nabla P$ took 7.2 sec. Using the modification above, each computation took only 3.5 sec. An additional benefit was increased accuracy, which resulted in fewer moves required to solve the problem. Using the packing techniques, only 603 memory locations were required to store $(\nabla^2 P)^{-1}$, compared to 2025 for the regular way.

Total time to solve this problem was 18 min. on the IBM 7040 using the revised method. The usual method—using the Newton second-order method in a straightforward way—was not run to completion, but based on its progress it would have taken about 90 min.

8.4 ACCELERATION BY EXTRAPOLATION

A very powerful computational tool is available when the conditions guaranteeing the existence of $D^i[x(0)]$ hold as developed in Chapter 5. In the following discussion these appropriate conditions are assumed to hold.

Suppose the W function has been uniquely minimized for $r_1 > \cdots > r_p > 0$ at $x^1, \ldots, x^p$. A polynomial in r that yields $x^1, \ldots, x^p$ is given by a set of equations of the form

$$x(r_k) = x^k = \sum_{j=0}^{p-1} a_j (r_k)^j, \qquad k = 1, \ldots, p, \tag{8.59}$$

where the a_j are n-component vectors.

The determinant of the matrix

$$R = \begin{bmatrix} r_1^0 & ,\ldots, & r_p^0 \\ \cdot & & \\ \cdot & & \\ \cdot & & \\ r_1^{p-1} & ,\ldots, & r_p^{p-1} \end{bmatrix}$$

(called the Vandermonde determinant [105, page 120]) is equal to

$$\prod_{i<j} (r_j - r_i)$$

and, since $r_j \neq r_i$ ($i \neq j$), R is nonsingular. Thus the vectors a_j are uniquely determined by (8.59). Then $\sum_{j=0}^{p-1} a_j(r)^j$ is an approximation of $x(r)$ in the interval $[0, r_1]$, and $x(0) = x^*$ (a solution) is approximated by a_0. That this approximation converges to a solution, and in fact that the estimate improves with each minimum that is determined, is seen as follows.

The exact Taylor series expansion of x^k in r_k about 0 is

$$x^k = \sum_{j=0}^{p-1} (r_k)^j \frac{D^j x(0)}{j!} + \epsilon^k, \qquad k = 1, \ldots, p, \tag{8.60}$$

where

$$\epsilon_k = \left(\frac{r_k^p}{p!}\right)\left[\frac{d^p x_1(\eta_{1k})}{dr^p}, \ldots, \frac{d^p x_n(\eta_{nk})}{dr^p}\right], \quad 0 \leq \eta_{jk} \leq r_k, j = 1, \ldots, n.$$

8.4 Acceleration by Extrapolation

Setting (8.59) and (8.60) equal, subtracting, and combining yields

$$[\epsilon^1, \ldots, \epsilon^p]R^{-1} = A - \left[x(0), \frac{D^1 x(0)}{1!}, \ldots, \frac{D^{p-1} x(0)}{(p-1)!} \right],$$

where $A = [a_0, \ldots, a_{p-1}]$. Clearly, then, the difference between a_0 and $x(0)$ is of the order of r_1^p. Thus, as $r_1 \to 0$, $a_0 \to x(0)$. More important, the estimates using p minima are better than those using $(p-1)$ minima. When $r_{k+1} = r_k/c$ ($c > 1$), the particular structure of these equations makes it possible to develop a simple iterative scheme for computing a series of estimates based on using a given number of terms in the polynomial. Notice that the a_j need not be computed to obtain these estimates.

Let $x_{i,j}$, $i = 1, \ldots, p$, $j = 0, \ldots, i-1$ signify the jth order estimate of $x(0)$ after i minima have been achieved, with the understanding that r_1 is the initial value of r. Note that the order corresponds to the index of a_j in (8.59). Then it follows that

$$x_{i,0} = x\left(\frac{r_1}{c^{i-1}}\right), \qquad i = 1, \ldots, p$$

and

$$x_{i,j} = \frac{(c^j) x_{i,j-1} - x_{i-1,j-1}}{c^j - 1}, \qquad i = 2, \ldots, p-1, \ j = 1, \ldots, i-1. \tag{8.61}$$

Then a_0, the "best" estimate of $x(0)$, is given by

$$x(0) \doteq x_{p,p-1} = a_0. \tag{8.62}$$

The extrapolation formulas (8.61) can also be used to estimate the next minimum of the W function after a number of minima have been computed. For example, the $(p+1)$st minimum, based on the information collected from the previous p minima, is estimated by

$$x_{p+1,0} = x\left(\frac{r_1}{c^p}\right) = a_0 + \cdots + a_{j-1}\left(\frac{r_1}{c^p}\right)^{j-1} + \cdots + a_{p-1}\left(\frac{r_1}{c^p}\right)^{p-1}. \tag{8.63}$$

Although the $\{a_j\}$ have not been computed explicitly, it is possible to make use of relations (8.61) and (8.62) and work backward to $x_{p+1,0}$. This is accomplished by setting $i = p + 1$ in (8.61) and solving for $x_{p+1,j-1}$. This gives the recursive relation

$$x_{p+1,j-1} = \frac{(c^j - 1) x_{p+1,j} + x_{p,j-1}}{c^j}. \tag{8.64}$$

Computational Aspects of Unconstrained Minimization Algorithms

Noting that $a_0 = x_{p,p-1} = x_{p+1,p-1}$ from (8.62) and using the values previously obtained from (8.61), we can evaluate (8.64) for $j = p-1, p-2, \ldots, 1$. The last computation will yield the required estimate $x_{p+1,0}$.

This estimate can be used as the starting point for the $(p+1)$st minimization of the W function. As more minima are achieved, the estimate eventually improves. This accelerates the entire process by substantially reducing the effort required to minimize the successive W functions.

In practice, computer storage requirements and accuracy considerations such as round-off error (which becomes critical for higher order estimates) limit the number of estimates possible. However, the authors' experience thus far indicates that considerable computational advantage is gained even when only first- and second-order estimates are made of a next W minimum and the optimum.

The following example shows how estimates can be used to accelerate convergence.

Example. The example of Figure 6, Section 4.3, is used to illustrate the extrapolation technique. In Table 10 are the data for the iterations.

Table 10 Use of Extrapolation Formulas to Accelerate Convergence

r_k	Estimates x_1^k (1)	(2)	(3)	Estimates x_2^k (1)	(2)	(3)	$\ln x_1 - x_2$ $f(x)$
1.0	1.5527902			1.3328244			−0.8927710
0.25	1.1593310			1.6413662			−1.4935231
		1.0281779			1.7442134		−1.7164252
6.25 × 10⁻²	1.0398432			1.7111091			−1.6720392
		1.0000139			1.7343567		−1.7343428
			0.9981363			1.7336995	−1.7355649
1.5625 × 10⁻²	1.0099207			1.7269415			−1.7170697
		0.9999465			1.7322189		−1.7322724
			0.9999420			1.7320763	−1.7321343
			0.9999706			1.7320505	−1.7320799
3.960625 × 10⁻³	1.0024774			1.7307819			−1.7283076
		0.99999630			1.7320620		−1.7320657
			0.99999960			1.7320515	−1.7320519
			1.0000005			1.7320511	−1.7320506
Theoretical solution	1.0			$\sqrt{3}$ $\doteq 1.7320505$			$-\sqrt{3}$ $\doteq -1.7320505$

The convergence of the estimates to the theoretical solution can be seen by reading down the columns. The power of the extrapolation formulas can be summed up by noting that the third-order estimates using the last four minima (last line before theoretical solution) agree with the theoretical solution to seven places, whereas the minimum for $r = 3.960625 \times 10^{-3}$ agrees to only three places.

8.5 OTHER ALGORITHMIC REQUIREMENTS

Parameter Selection

In Section 7.1 ways of choosing initial weighting factors for each term in the unconstrained function were given. Here it is assumed that the same value of r is applied to each term.

Since a necessary condition for minimization of $W(x, r)$ is the vanishing of the first partial derivatives, a natural choice of r_1 would be given by the r that minimizes the magnitude of the gradient at x^0; that is,

$$|\nabla W(x^0, r_1)| = \min_{r \geq 0} |\nabla W(x^0, r)|. \tag{8.65}$$

This criterion can be used provided that it yields $r_1 > 0$. If $r_1 = 0$, then progress can be made in the minimization of both $f(x)$ and the penalty terms. In this case $f(x)$ can be treated as an unconstrained minimization until some boundary defined by $g_i(x) > 0$, $i = 1, \ldots, m$, is violated. Recomputing (8.65) at a feasible point near the boundary must yield $r_1 > 0$. In general, of course, $r_1 > 0$ can be arbitrarily selected.

Within very wide tolerances, the factor c by which each r_k is reduced is not crucial to the convergence of the algorithm. The reasons for this are twofold. First, for any fixed c, more iterations are generally required to minimize the function at the next iteration than would be required if c were smaller. Second, the extrapolation techniques of Section 8.5 avoid the necessity for calculating unconstrained minima for small values of r.

Minimizing the W Function along a Vector

It is assumed that the W function has finite minima for every $r > 0$. Conditions sufficient to ensure this were given in Sections 4.3, 6.3, and 6.4. After a direction in which the W function decreases has been computed, it is necessary to decide on the step size. If s is the direction in which the move is to be made (it may be the negative gradient of W, or the second-order gradient, or some conjugate direction given by the other methods), then the subject of this section is how to choose and compute the scalar θ, where

$$x^2 = x^1 + \theta s. \tag{8.66}$$

The method that has worked best has been to choose θ such that

$$W(x^2, r) = \min_{\theta \geq 0} W(x^1 + \theta s, r), (x^1 + \theta s \in R^0).$$

This is a one-dimensional minimization problem. When W is strictly convex, any θ that accomplishes this yields a unique minimum value (that is, a local minimum in θ is a global minimum), since the function W is unimodal in θ. We shall assume that W is convex. (A slight modification must be made for nonconvex functions.)

It should be noted that the problem of selecting θ arises in all methods of mathematical programming, in various forms. The method of feasible directions, the projected gradient technique, and the simplex method require further that all points remain feasible. Since the W function increases without bound as the boundary of the feasible region is approached, the minimization here remains unconstrained and necessarily confined to the interior of the feasible region.

Perhaps the most straightforward technique for determining a suitable value of θ, say $\bar{\theta}$, is a simple bracketing procedure whereby upper and lower bounds on $\bar{\theta}$ are initially computed. An intermediate value of $\bar{\theta}$ is then tested and used to generate information to sharpen the bounds. The process is repeated until the bounds are satisfactorily tight.

Following is a description of the bracketing technique used in an early computer implementation of the algorithm. (The method is quite satisfactory for small problems.) Assume $\nabla W(x^1, r) \neq 0$; otherwise x^1 is the required minimum of W. In order to determine how to adjust any particular choice of θ, it is necessary to know whether the W function is increasing or decreasing at any point on s. If W is a function of θ it follows that

$$\frac{dW(x^1 + \theta s, r)}{d\theta} = \nabla W(x^1 + \theta s, r)^T \left(\frac{dx}{d\theta}\right) = \nabla W(x^1 + \theta s, r)^T s.$$

This rate of change of W at $\theta = \theta_i$ will be designated $W_\theta(\theta_i)$, $i = 1, 2, \ldots$. Let $\theta_0 = 0$.

From the above assumptions it follows that $s \neq 0$, and $W_\theta(\theta_0) < 0$. The following steps are then taken.

STEP 1. Compute $W_\theta(\theta_i)$ for $i = 1, 2, 3, \ldots$, where $\{\theta_i\}$ is any strictly monotonic increasing unbounded sequence of positive numbers such that $(x + \theta_i s) \in R$ for all i. If $W_\theta(\theta_i) \doteq 0$ for some $i = N$, then θ is the required approximation of $\bar{\theta}$. Otherwise let M be the smallest value of i for which $W_\theta(\theta_i) \geq 0$. (Otherwise W has unbounded minima, contrary to assumption.) It follows that $W_\theta(\theta_{M-1}) \leq 0$, so that $\theta_L \leq \bar{\theta} \leq \theta_U$, where $\theta_L = \theta_{M-1}$ and $\theta_U = \theta_M$.

8.5 Other Algorithmic Requirements

STEP 2. Compute $W_\theta[(\theta_U + \theta_L)/2] = W_\theta(\theta^*)$. If $W_\theta(\theta^*) \doteq 0$, $\bar\theta \doteq \theta^*$. If $W_\theta(\theta^*) < 0$, the new lower bound on $\bar\theta$ is given by $\theta'_L = \theta^*$. If $W_\theta(\theta^*) > 0$, the new upper bound is $\theta'_U = \theta^*$.

STEP 3. Repeat Step 2, replacing θ_U by θ'_U and θ_L by θ'_L. Terminate when $\theta'_U - \theta'_L < \epsilon$, a specified positive constant. The final approximation of $\bar\theta$ is given by $(\theta'_U + \theta'_L)/2$.

The accuracy of the problem solutions when this method of minimizing was used was very satisfactory. However, analysis of the computer results showed that the gradient computations required an inordinate fraction of the over-all computational effort.

Another method of estimating θ that involved only the *values* of W at successive trial points was then tried. It is based on an optimal search procedure in one dimension utilizing the properties of the Fibonacci series [119]. The following steps define the procedure.

STEP 1. First an upper bound θ_U is obtained on $\bar\theta$. (For the first lower bound, $\theta_L = 0$.) θ_U is obtained by evaluating the W function at successive points whose θ values are in the limiting Fibonacci ratio, $1.618 \doteq (1 + \sqrt{5})/2$; that is,

$$\theta_U = \sum_{i=0}^{\bar i} (1.618)^i,$$

where $\bar i$ is the smallest non-negative integer i such that

$$W\left[x^1 + \sum_{j=0}^{i}(1.618)^j s\right] > W[x^1 + \sum_{j=0}^{i-1}(1.618)^j s]$$

STEP 2. The interval bounding $\bar\theta$ is reduced by computing two values of θ in the interval (θ_U, θ_L),

$$\theta_a = \theta_L + 0.382(\theta_U - \theta_L)$$

and

$$\theta_b = \theta_L + 0.618(\theta_U - \theta_L).$$

STEP 3. The W values of the (interior) points corresponding to θ_a and θ_b are compared.

STEP 4. If $W(x^1 + \theta_a s, r) < W(x^1 + \theta_b s, r)$, then $\theta_L \leq \bar\theta \leq \theta_b$. Because of the property that $0.382/0.618 \doteq 0.618$, by letting $\theta'_U = \theta_b$, $\theta'_b = \theta_a$, and $\theta'_L = \theta_L$, and recomputing θ'_a as $\theta'_a = \theta'_L + 0.382(\theta'_U - \theta'_L)$, Step 3 can now be repeated.

STEP 5. If $W(x^1 + \theta_a a, r) > W(x^1 + \theta_b s, r)$, then by letting $\theta'_L = \theta_a$, $\theta'_a = \theta_b$, and $\theta'_U = \theta_U$, and computing $\theta'_b = \theta'_L + 0.618(\theta'_U - \theta'_L)$, Step 3 can be repeated.

STEP 6. If $W(x^1 + \theta_a s, r) = W(x^1 + \theta_b s, r)$, let $\theta'_L = \theta_a$, $\theta'_U = \theta_b$, and repeat Step 2.

STEP 7. When $\theta_U - \theta_L$ is acceptably small, $\bar\theta$ is approximated by $\bar\theta = (\theta_L + \theta_U)/2$, and $x^2 = x^1 + \bar\theta s$.

Modifications of this procedure are made in Step 1 when a trial point is not in the interior of the feasible region. In this case the W value there is considered to have infinite value and the corresponding θ is an upper bound on $\bar\theta$. Also, in Step 2 computational advantage is realized when one or more interior points have been computed in Step 1. (For nonconvex functions, if the value of W at the left interior point is greater than the left endpoint value, that interior point immediately becomes the right endpoint and the process is restarted from Step 2.)

Other methods for minimizing a function along a vector have been proposed in [35] and [103].

Convergence Criteria

The smallest value of r for which an unconstrained minimum is to be found depends in an essential way on the convergence criteria specified by the user.

Let $\hat x$ be a point to be tested for optimality in Problem M. We assume here that (M) is a convex programming problem. For nonconvex problems no lower bound is available on the optimum value. (In practice the same criterion given in the following is used for the latter problems, but without the attendant theoretical justification.) The point $\hat x$ may have been obtained from the extrapolation formulas (8.61–8.64), or it may be a minimum of W for some value of r. First, it must be decided if $\hat x$ is acceptably feasible, that is, if

$$\max\,[-g_i(\hat x), \quad i = 1, \ldots, m; \quad |g_{m+j}(\hat x)|, \quad j = 1, \ldots, q - m] \le \delta,$$

where $\delta > 0$ is the acceptable amount by which a constraint may be violated (according to the specification of the user). If $\hat x$ is acceptably feasible, then the duality theory of Section 6.2 leads to a check on how close $f(\hat x)$ is to the optimum v^*.

For the W function the dual objective function value is given by

$$G^k = f(x^k) - mr_k + \frac{2}{r_k} \sum_{i=m+1}^{q} \{\min\,[0, g_i(x^k)]\}^2 \le v^*,$$

where r_k is the smallest value of r for which $W(x, r)$ has been minimized. Thus, if

$$f(\hat x) - G^k < \epsilon, \tag{8.67}$$

it follows that $f(\hat x) < v^* + \epsilon$.

When the *magnitude* of the tolerable error in estimating v^* cannot be specified in advance, the following provides a normalized criterion. Define the "fractional error" as the ratio of the absolute error to the solution value. In terms of the above notation this is given by $[f(\hat{x}) - v^*]/v^*$. Now, assuming $v^* \neq 0$, $G^k \neq 0$, and that these values have the same sign, it follows that

$$\frac{f(\hat{x}) - G^k}{G^k} \geq \frac{f(\hat{x}) - v^*}{v^*} \geq 0. \tag{8.68}$$

Thus, if the left side of inequality (8.68) is less than 10^{-i} it may be verified in general that $f(\hat{x})$ agrees with v^* in approximately i places.

When v^* is close to 0, the criterion furnished by (8.68) may not be too reliable since all the terms in the ratio may be small, but in this case (8.67) provides a satisfactory criterion. However, in the absence of further information about the problem, (8.68) generally provides a satisfactory convergence criterion.

Finding an Interior Point

If a point interior to the constraints, $g_i(x) \geq 0$, $i = 1, \ldots, m$, is initially unavailable, repeated application of the method itself can be used to get such a point. A description of a procedure for minimizing the P function was given by Fiacco [47] to solve an auxiliary problem for each violated constraint. A modification of this, where an attempt is made to satisfy all violated constraints at once, has proved more efficient computationally and is reviewed here.

The problem is to determine a point that satisfies the strict inequalities, $g_i(x) > 0$, $i = 1, \ldots, m$, as required by interior point algorithms. Assume that x^0 is given, but that not all of the inequalities are satisfied. Define the sets $S = \{s \mid g_s(x) \leq 0, 1 \leq s \leq m\}$ and $T = \{t \mid g_t(x) > 0, 1 \leq t \leq m\}$. The strategy is to select a sequence of points that increases the value of $\sum_{s \in S} g_s(x)$ without violating any of the conditions already satisfied; that is, the $g_t(x) > 0$. This can be done by maximizing $\sum_{s \in S} g_s(x)$ [or equivalently by minimizing $-\sum_{s \in S} g_s(x)$] by minimizing

$$U(x, r_k) \equiv -\sum_{s \in S} g_s(x) + r_k \sum_{t \in T} I_t[g_t(x)]$$

for a strictly decreasing (null) sequence $\{r_k\}$.

This is called the auxiliary problem. [The constraints $(g_{m+1}, \ldots, g_q)$ to be handled by exterior point penalty terms are ignored until an interior point is obtained.] Each time a new point in the process strictly satisfies one or more of the previously violated constraints, the corresponding indices are removed from S and placed in T, and the auxiliary problem is modified accordingly. Finally, when S is empty, an interior point is attained. If at a solution of the auxiliary problem S is not empty, the problem is considered infeasible.

If the constraints are concave, the auxiliary problem is a convex programming problem. Because of this, an easy check on the nonfeasibility of the problem is available at any unconstrained minimum. Using the inequalities resulting from the dual theory (Section 6.2), it follows that if $G^k > 0$ at the kth minimum the problem is not feasible. Even though the current auxiliary problem may be far from being solved, further computations are not needed to determine nonfeasibility. This is a distinct advantage of the dual aspects of this approach.

REFERENCES

1. Abadie, J., J. Carpentier, and C. Hensgen, "Generalization of the Wolfe Reduced Gradient Method to the Case of Nonlinear Constraints," paper presented at the Joint European Meeting of the Econometric Society/The Institute of Management Science, Warsaw, September 1966.
2. Ablow, C. M., and G. Brigham, "An Analog Solution of Programming Problems," *Operations Res.*, **3**:388–394 (1955).
3. Apostol. T. M., *Mathematical Analysis*, Addison-Wesley, Reading, Mass., 1957.
4. Arrow, K. J., "A Gradient Method for Approximating Saddle Points and Constrained Maxima," RAND Corp., Santa Monica, Calif., Paper P-223, 15 pp., 1951.
5. Arrow, K., and L. Hurwicz, "Reduction of Constrained Maxima to Saddle Point Problems," in J. Neyman (Ed.), *Proceedings of the Third Berkeley Symposium on Mathematical Statistics and Probability*, University of California Press, Berkeley, 1956, pp. 1–20.
6. Arrow, K. J., L. Hurwicz, and H. Uzawa, *Studies in Linear and Nonlinear Programming*, Stanford University Press, Palo Alto, Calif., 1958.
7. Beale, E. M. L., "On Minimizing a Convex Function Subject to Linear Inequalities," *J. Roy. Statist. Soc., Ser. B*, **17**(2):173–184 (1955).
8. Beale, E. M. L., "On Quadratic Programming," *Naval Res. Logist. Quart.*, **6**:227–243 (September 1959).
9. Beltrami, E. J., and R. McGill, "A Class of Variational Problems in Search Theory and the Maximum Principle," *Operations Res.*, **14**(2):267–278 (1966).
10. Berge, C., and A. Ghouila-Houri, *Programmes, Games and Transportation Networks*, Wiley, New York, 1965.
11. Berkovitz, L. D., "Variational Methods in Problems of Control and Programming," *J. Math. Anal. Appl.*, **3**:145–169 (1961).
12. Bochner, S., and W. T. Martin, *Several Complex Variables*, Princeton University Press, Princeton, N.J., 1948.
13. Bombart, J., and F. C. Hipp, "Analogue Computer Solution of Programming Problems," Memorandum, Research Analysis Corp., McLean, Virginia.
14. Bracken, J., and G. P. McCormick, *Selected Applications of Nonlinear Programming*, Wiley, New York, 1968.
15. Broyden, C. G., "A Class of Methods for Solving Nonlinear Simultaneous Equations," *Math. Computation*, **19**(92):577–593 (1965).

References

16. Bui Trong Lieu and P. Huard, "La Méthode des Centres dans un Espace Topologique," *Numér. Math.*, **3**(1):56–67 (1966).
17. Butler, T., and A. V. Martin, "On a Method of Courant for Minimizing Functionals," *J. Math. Phys.*, **41**:291–299 (1962).
18. Camp, G. D., "Inequality-Constrained Stationary Value Problems," *J. Operations Res. Soc. Am.*, **3**:548–550 (1955).
19. Carathéodory, C., *Variatonsrechnung und Partielle Differentialgleichungen erste Ordung*, Teubner, Leipzig, 1935.
20. Carroll, C. W., *An Operations Research Approach to the Economic Optimization of a Kraft Pulping Process*, Ph.D. Dissertation, Institute of Paper Chemistry, Appleton, Wis., 1959.
21. Carroll, C. W., "The Created Response Surface Technique for Optimizing Nonlinear Restrained Systems," *Operations Res.*, **9**(2):169–184 (1961).
22. Cauchy, A., "Méthode Générale pour la Resolution des Systèmes D'équations Simultanées," *Compt. Rend.*, **25**:536–538 (1847).
23. Charnes, A., and W. W. Cooper, Letter to the Editor, "Such Solutions are Very Little Solved," *Operations Res.*, **3**(3):345–346 (1955).
24. Charnes, A., and W. W. Cooper, "Chance-Constrained Programming," *Management Sci.*, **6**:73–79 (1959).
25. Charnes, A., and W. W. Cooper, *Management Models and Industrial Applications of Linear Programming*, Wiley, New York, 1961.
26. Charnes, A., W. W. Cooper, and A. Henderson, *An Introduction to Linear Programming*, Wiley, New York, 1953.
27. Charnes, A., and C. E. Lemke, "Minimization of Non-linear Separable Convex Functionals," *Naval Res. Logist. Quart.*, **1**:301–302 (1954).
28. Cheney, E. W., and A. A. Goldstein, "Newton's Method for Convex Programming and Tchebycheff Approximation," *Numer. Math.*, **1**:253–268 (1959).
29. Courant, R., "Variational Methods for the Solution of Problems of Equilibrium and Vibrations," *Bull. Am. Math. Soc.*, **49**:1–23 (1943).
30. Courant, R., "Calculus of Variations and Supplementary Notes and Exercises" (Mimeographed Notes), Supplementary Notes by M. Kruskal and H. Rubin, revised and amended by J. Moser, New York University, 1956–1957.
31. Crockett, J. B., and H. Chernoff, "Gradient Methods of Maximization," *Pacific J. Math.*, **5**:33–50 (1955).
32. Curry, H. B., "The Method of Steepest Descent for Non-linear Minimization Problems," *Quart. Appl. Math.*, **2**(3):258–261 (1944).
33. Dantzig, G. B., "Maximization of a Linear Function of Variables Subject to Linear Inequalities," in T. C. Koopmans (Ed.), *Activity Analysis of Production and Allocation*, Wiley, New York, 1951, Chapter 21.
34. Dantzig, G., *Linear Programming and Extensions*, Princeton University Press, Princeton, N.J., 1963.
35. Davidon, W. C., "Variable Metric Method for Minimization," Research and Development Report ANL-5990 (Revised), Argonne National Laboratory, U.S. Atomic Energy Commission, 1959.
36. DeHaven, J. C., and E. C. Deland, "Reactions of Hemoglobin and Steady States in the Human Respiratory System: An Investigation Using Mathematical Models and an Electronic Computer," RAND Corp., Santa Monica, Calif., Memorandum RM-3212-PR, December 1962, pp. 174–175.
37. Dorn, W. S., "On Lagrange Multipliers and Inequalities," *J. Operations Res. Soc. Am.*, **9**:95–104 (1961).

38. Duffin, R. J., "Cost Minimization Problems Treated by Geometric Means," *Operations Res.*, **10**:668–675 (1962).
39. Duffin, R. J., E. L. Peterson, and C. Zener, *Geometric Programming—Theory and Application*, Wiley, New York, 1967.
40. Everett, H., "Generalized Lagrange Multiplier Method for Solving Problems of Optimum Allocation of Resources," *Operations Res.*, **11**:399–417 (1963).
41. Falk, J. E., "A Constrained Lagrangian Approach to Nonlinear Programming," Ph.D. Dissertation, University of Michigan, Ann Arbor, June 1965.
42. Falk, J. E., "A Relaxed Interior Approach to Nonlinear Programming," Technical Paper RAC-TP-279, Research Analysis Corporation, McLean, Va., 1967.
43. Falk, J. E., "Lagrange Multipliers and Nonlinear Programming," *J. Math. Anal. Appl.*, **19**(1) (July 1967).
44. Farkas, J., "Uber die Theorie der einfachen Ungleichungen," *J. Reine Angew. Math.*, **124**:1–27 (1901).
45. Faure, P., and P. Huard, "Résolution des Programmes Mathématiques à Fonction Non-linéaire par la Méthode du Gradient Reduit," *Rev. Franc. Recherche Opérationelle*, **9**:167–205 (1965).
46. Feder, D. P., "Automatic Lens Design Methods," *J. Opt. Soc. Am.*, **47**:902–912 (1957).
47. Fiacco, A. V., "Comments on the Paper of C. W. Carroll," *Operations Res.*, **9**:184–185 (1961).
48. Fiacco, A. V., "Sequential Unconstrained Minimization Methods for Nonlinear Programming," Ph.D. Dissertation, Northwestern University, Evanston, Ill., June 1967.
49. Fiacco, A. V., and G. P. McCormick, "Programming under Nonlinear Constraints by Unconstrained Minimization: A Primal-Dual Method," Technical Paper RAC-TP-96, Research Analysis Corporation, McLean, Va., 1963.
50. Fiacco, A. V., and G. P. McCormick, "The Sequential Unconstrained Minimization Technique for Nonlinear Programming: A Primal-Dual Method," *Management Sci.*, **10**(2):360–366 (1964).
51. Fiacco, A. V., and G. P. McCormick, "Computational Algorithm for the Sequential Unconstrained Minimization Technique for Nonlinear Programming," *Management Sci.*, **10**(4):601–617 (1964).
52. Fiacco, A. V., and G. P. McCormick, "SUMT without Parameters", System Research Memorandum No. 121, Technical Institute, Northwestern University, Evanston, Ill., 1965.
53. Fiacco, A. V., and G. P. McCormick, "Extensions of SUMT for Nonlinear Programming: Equality Constraints and Extrapolation," *Management Sci.*, **12**(11):816–829 (1966).
54. Fiacco, A. V., and G. P. McCormick, "The Slacked Unconstrained Minimization Technique for Convex Programming," *SIAM J. Appl. Math.*, **15**(3):505–515 (1967).
55. Fletcher, R., "Function Minimization without Evaluating Derivatives—A Review," *Computer J.*, **8**:33–41 (1965).
56. Fletcher, R., and M. J. D. Powell, "A Rapidly Convergent Descent Method for Minimization," *Computer J.*, **6**:163–168 (1963).
57. Fletcher, R., and C. M. Reeves, "Function Minimization by Conjugate Gradients," *Computer J.*, **7**:149:154 (1964).
58. Flood, M. M., and A. Leon, "A Generalized Search Code for Optimization," Preprint 129, Mental Health Research Institute, University of Michigan, Ann Arbor, 1964.

59. Frisch, K. R., "Principles of Linear Programming—With Particular Reference to the Double Gradient Form of the Logarithmic Potential Method," Memorandum of October 18, 1954, University Institute of Economics, Oslo.
60. Frisch, K. R., "The Logarithmic Potential Method of Convex Programming," Memorandum of May 13, 1955, University Institute of Economics, Oslo.
61. Gale, D., "Convex Polyhedral Cones and Linear Inequalities," in T. C. Koopmans (Ed.), *Activity Analysis of Production and Allocation*, Wiley, New York, 1951.
62. Gass, S. I., *Linear Programming: Methods and Applications*, McGraw-Hill, New York, 1958.
63. Gelfand, I. M., and S. V. Fomin, *Calculus of Variations*, Revised English Edition, Translated and Edited by R. A. Silverman, Prentice-Hall, Englewood Cliffs, N.J., 1963, Chapter 2.
64. Geoffrion, A. M., "Strictly Concave Parametric Programming, Part I: Basic Theory," *Management Sci.*, 13:244–253 (1966).
65. Geoffrion, A. M., "Strictly Concave Parametric Programming, Part II: Additional Theory and Computational Considerations," *Management Sci.*, 13(5):359–370 (1967).
66. Goldstein, A. A., "Cauchy's Method of Minimization," *Numer. Math.*, 4:146–150 (1962).
67. Goldstein, A. A., and B. R. Kripke, "Mathematical Programming by Minimizing Differentiable Functions," *Numer. Math.*, 6:47–48 (1964).
68. Hadley, G., *Nonlinear and Dynamic Programming*, Addison-Wesley, Reading, Mass., 1964.
69. Hancock, H., *Theory of Maxima and Minima*, Dover, New York, 1960.
70. Hartley, H. O., and R. R. Hocking, "Convex Programming by Tangential Approximation," *Management Sci.*, 9:600–612 (1963).
71. Hestenes, M. R., *Calculus of Variations and Optimal Control Theory*, Wiley, New York, 1966.
72. Householder, A. S., *The Theory of Matrices in Numerical Analysis*, Blaisdell, New York, 1964.
73. Huard, P., "Résolution des P. M. a Constraintes Non-lineaires par la Méthode des Centres," Note E.D.F. HR 5.690, May 6, 1964.
74. Kelley, J. E., Jr. "The Cutting Plane Method for Solving Convex Programs," *J. Soc. Ind. Appl. Math.*, 8(4):703–712 (1960).
75. Kiefer, J., "Optimum Sequential Search and Approximation Methods under Minimum Regularity Assumptions," *SIAM J.*, 5:105–136 (1957).
76. Klein, B., "Direct Use of Extremal Principles in Solving Certain Optimizing Problems Involving Inequalities," *J. Operations Res. Soc. Am.*, 3:548 (1955).
77. Kuhn, H. W., and A. W. Tucker, "Non-linear Programming," in J. Neyman (Ed.), *Proceedings of the Second Berkeley Symposium on Mathematical Statistics and Probability*, University of California Press, Berkeley, 1951, pp. 481–493.
78. Leitmann, G. (Ed.), *Optimization Techniques: With Applications to Aerospace Systems*, Academic Press, New York, 1962.
79. Lemke, C. E., "A Method for Solution of Quadratic Programming Problems," *Management Sci.*, 8:442–453 (1962).
80. Leon, A., "A Comparison among Eight Known Optimizing Procedures," in A. Lavi and T. P. Vogl (Eds.), *Proceedings of Symposium on Recent Advances in Optimization Techniques*, Wiley, New York, 1965.
81. Leon, A., "An Annotated Bibliography on Optimization," Preprint 162, Mental Health Research Institute, University of Michigan, Ann Arbor, 1965.
82. Mangasarian, O. L., and S. Fromowitz, "The Fritz John Necessary Optimality

Conditions in the Presence of Equality and Inequality Constraints," *J. Math. Anal. Appl.*, **17**(1):37–47 (1967).
83. Markowitz, H., *Portfolio Selection*, Wiley, New York, 1959.
84. McCormick, G. P., "Second Order Conditions for Constrained Minima," *SIAM J. Appl. Math.*, **15**(3):641–652 (1967).
85. McCormick, G. P., W. C. Mylander, III, and A. V. Fiacco, "Computer Program Implementing the Sequential Unconstrained Minimization Technique for Nonlinear Programming," Technical Paper RAC-TP-151, Research Analysis, McLean, Va., 1965.
86. McCormick, G. P., and W. I. Zangwill, "A Technique for Calculating Second Order Optima," Technical Paper, Research Analysis Corp., McLean, Virginia.
87. McGill, R., and P. Kenneth, "Solution of Variational Problems by Means of a Generalized Newton-Raphson Operator," *AIAA J.*, **2**(10):1761–1766 (1964).
88. Miller, C. E., "The Simplex Method for Local Separable Programming," in R. L. Graves and P. Wolfe (Eds.), *Recent Advances in Mathematical Programming*, McGraw-Hill, New York, 1963, pp. 89–100.
89. Motzkin, T. S., "New Techniques for Linear Inequalities and Optimization," in *Project SCOOP, Symposium on Linear Inequalities and Programming*, Planning Research Division, Director of Management Analysis Service, U.S. Air Force, Washington, D.C., No. 10, 1952.
90. Murray, W., "Ill-Conditioning in Barrier and Penalty Functions Arising in Constrained Non-linear Programming," paper presented at the Princeton Mathematical Programming Symposium, August 14–18, 1967.
91. Parisot, G. R., "Résolution Numérique Approchée du Problème de Programmation Lineaire par Application de la Programmation Logarithmique," Ph.D. Dissertation, University of Lille, France, 1961.
92. Pietrzykowski, T., "Application of the Steepest Descent Method to Concave Programming," in *Proceedings of the IFIPS Congress*, Munich, 1962, North-Holland Publishing Company, Amsterdam, 1962, pp. 185–89.
93. Pomentale, T., "A New Method for Solving Conditioned Maxima Problems," *J. Math. Anal. Appl.*, **10**:216–220 (1965).
94. Pontryagin, L. S., V. G. Boltyanskii, R. V. Gamkrelidze, and E. F. Mishchenko, *The Mathematical Theory of Optimal Processes*, Translated by K. N. Trinogoff, Interscience, New York, 1962.
95. Powell, M. J. D., "An Efficient Method for Finding the Minimum of a Function of Several Variables without Calculating Derivatives," *Computer J.*, **7**(2):155–162 (1964).
96. Riley, V., and S. I. Gass, *Linear Programming and Associated Techniques*, Johns Hopkins Press, Baltimore, 1958.
97. Ritter, K., "Stationary Points of Quadratic Maximum Problems," *Z. Wahrscheinlichkeits—Theorie verwandte Gebeite*, **4**:149–158 (1965).
98. Rosen, J. B., "The Gradient Projection Method for Nonlinear Programming, Part I: Linear Constraints," *J. Soc. Ind. Appl. Math.*, **8**(1):181–217 (1960).
99. Rosen, J. B., "The Gradient Projection Method for Nonlinear Programming, Part II: Nonlinear Constraints," *J. Soc. Ind. Appl. Math.*, **9**:514–532 (1961).
100. Rosenbrock, H. H., "Automatic Method for Finding the Greatest or Least Value of a Function," *Computer J.*, **3**:175–184 (1960).
101. Rubin, H., and P. Ungar, "Motion under a Strong Constraining Force," *Commun. Pure Appl. Math.*, **10**:65–87 (1957).
102. Rudin, W., *Principles of Mathematical Analysis*, McGraw-Hill, New York, 1953.
103. Spang, H. A., "A Review of Minimization Techniques for Nonlinear Functions," *SIAM Rev.*, **4**(4):343–365 (1962).

104. Stong, R. E., "A Note on the Sequential Unconstrained Minimization Technique for Non-linear Programming," *Management Sci.*, **12**(1):142–144 (1965).
105. Thrall, R. M., and L. Tornheim, *Vector Spaces and Matrices*, Wiley, New York, 1957.
106. Valentine, F. A., "The Problem of Lagrange with Differential Inequalities as Added Side Conditions," in *Contributions to the Calculus of Variations, 1933–1937*, University of Chicago Press, Chicago, 1937, pp. 407–448.
107. White, W. B., S. H. Johnson, and G. B. Dantzig, "Chemical Equilibrium in Complex Mixtures," *J. Chem. Phys.*, **28**:751–755 (1958).
108. Widder, D. V., *Advanced Calculus*, 2nd Ed., Prentice-Hall, Englewood Cliffs, N.J., 1961, Chapter 4.
109. Wilde, D. J., *Optimum Seeking Methods*, Prentice-Hall, Englewood Cliffs, N.J., 1964.
110. Wolfe, P., "The Simplex Method for Quadratic Programming," *Econometrica*, **27**(3):382–398 (1959).
111. Wolfe, P., "A Duality Theorem for Nonlinear Programming," *Quart. Appl. Math.*, **19**(3):239–244 (1961).
112. Wolfe, P., "Methods of Nonlinear Programming," in R. L. Graves and P. Wolfe (Eds.), *Recent Advances in Mathematical Programming*, McGraw-Hill, New York, 1963, pp. 67–86.
113. Wolfe, P., "Foundations of Nonlinear Programming: Notes on Linear Programming and Extensions—Part 65," Memo RM-4669-PR, RAND Corp., Santa Monica, Calif., 1965.
114. Woodbury, M., "Inverting Modified Matrices," Memorandum Report 42, Statistical Research Group, Princeton, N.J., 1950.
115. Zangwill, W. I., "Non-linear Programming by Sequential Unconstrained Maximization," Working Paper 131, Center for Research in Management Science, University of California, Berkeley, 1965.
116. Zangwill, W. I., "On Minimizing a Function without Calculating Derivatives," Working Paper 210, Center for Research in Management Science, University of California, Berkeley, March 1967.
117. Zangwill, W. I., "Non-linear Programming via Penalty Functions," *Management Sci.*, **13**(5):344–358 (1967).
118. Zoutendijk, G., *Methods of Feasible Directions*, Elsevier, Amsterdam, 1960.
119. Zoutendijk, G., "Nonlinear Programming: A Numerical Survey," *SIAM J. Control*, **4**:194–210 (1966).

Index of Theorems, Lemmas, and Corollaries

Lemma 1 [Farkas (44)], 18
Theorem 1 [Existence of Generalized Lagrange Multipliers], 19
Corollary 1 [Kuhn-Tucker Necessity Theorem], 20
Corollary 2 [Interiority-Independence Necessity Theorem], 21
Corollary 3 [Sufficient Condition for First-Order Constraint Qualification], 22
Corollary 4 [First-Order Necessary Conditions for an Unconstrained Minimum], 24
Theorem 2 [Second-Order Necessary Conditions], 25
Theorem 3 [Condition Sufficient for Second-Order Constraint Qualification], 26
Corollary 5, 28
Theorem 4 [Second-Order Sufficiency Conditions], 30
Corollary 6, 31
Corollary 7 [Jacobian Condition Implying Sufficiency], 32
Theorem 5 [Neighborhood Sufficiency Theorem], 33
Theorem 6 [Perturbation of Optimum], 34
Lemma 2, 46
Corollary 8, 46
Theorem 7 [Existence of Compact Perturbation Set], 47
Theorem 8 [Convergence to Compact Sets of Local Minima by Interior Point Algorithms], 47
Corollary 9, 50
Corollary 10, 50
Theorem 9 [Convergence to Compact Sets of Local Minima by Exterior Point Algorithms], 57
Corollary 11, 59
Corollary 12, 59
Theorem 10 [Convergence of Mixed Algorithms], 60
Corollary 13, 61
Corollary 14, 61
Theorem 11 [Convergence of V Minima to Local Solutions of Problem M], 66
Corollary 15 [Convergence of U Minima to Local Solutions of Problem M(R)], 67
Corollary 16 [Convergence of T Minima to Local Solutions of Problem M(Q)], 68
Lemma 3, 68
Lemma 4, 69
Lemma 5, 69
Corollary 17, 69
Lemma 6, 70
Lemma 7, 71
Theorem 12 [Existence of an Isolated Trajectory], 73
Theorem 13 [Existence of $D^k x(r)$ for $r > 0$], 77
Corollary 18 [Analyticity of $x(r)$ for $r > 0$], 78
Theorem 14 [Existence of $Dx(0), Du(0)$], 80
Theorem 15 [Existence of $D^k x(r), D^k u(r)$], 81

204   Index of Theorems, Lemmas, and Corollaries

Corollary 19 [Analyticity of $x(r)$ at $r = 0$], 82
Theorem 16 [Existence of Isolated Trajectory for Quadratic Loss Function], 82
Theorem 17 [Isolated Trajectory for W Function], 84
Lemma 8, 87
Lemma 9, 88
Lemma 10, 88
Theorem 18 [Local-Global Convexity Property], 89
Theorem 19 [Kuhn-Tucker Sufficiency Theorem], 90
Theorem 20 [Nondifferentiable Sufficiency Theorem], 91
Theorem 21 [Nondifferentiable Necessary Conditions], 91
Theorem 22 [Primal-Dual Bounds], 92
Theorem 23 [Existence of Solution to Dual], 93
Theorem 24 [Preservation of Boundedness of Convex Sets Given by Concave Inequalities], 93
Corollary 20, 94
Lemma 11 [Convexity of the Unconstrained Function], 94
Lemma 12 [Boundedness of U Contours], 95
Theorem 25 [Convergence to Solution of Primal Convex Programming Problem by Interior Point Algorithms], 97
Theorem 26 [Dual Convergence-Interior Point Algorithms], 98
Lemma 13 [Convexity of T Function], 102
Lemma 14 [Existence of T Minima], 103
Theorem 27 [Solution of Convex Problems by Exterior Point Algorithms], 104
Theorem 28 [Exterior Point Dual Convergence], 105
Theorem 29 [Upper Bounds for v^*], 107
Lemma 15, 108
Theorem 30 [Strict Convexity of Analytic Penalty Function], 110
Theorem 31 [Existence of Basis for Linear Problem], 111
Corollary 21, 112
Theorem 32 [Convergence of Q-Function Algorithms to Compact Sets of Local Solutions], 122
Theorem 33 [Stability of the Stationary Points of the Lagrangian], 126
Theorem 34 [Global Stability for the Convex Programming Problem], 127
Theorem 35 [Solution of Strictly Convex Problem by Dual Lagrange Multiplier Algorithm], 131
Theorem 36 [Differentiability of $\gamma(u)$], 133
Theorem 37 [Convergence of ϵ-Lagrange Multiplier Approach to Solution of Convex Programming Problem], 136
Theorem 38 [Optimality of Lagrangian Maximum for Corresponding Constrained Problem], 139
Theorem 39 [Bounds on Ratio of Increased Payoff to Increased Resource Allocation Using Multipliers], 140
Theorem 40 [Existence of t^0 for Which Unconstrained Minimum Is Also a Constrained Minimum], 143
Theorem 41 [Convergence in n Steps for a Positive Definite Quadratic Form], 171
Theorem 42 [Use of Second Derivative Information], 173
Lemma 16, 174

Author Index

Ablow, C. M., 7, 143
Arrow, K. J., 6, 9

Beltrami, E. J., 14
Bochner, S., 78
Bombart, J., 128
Brigham, G., 7, 143
Bui Trong Lieu, 13, 121
Butler, T., 11

Camp, G. D., 8
Carathéodory, C., 38
Carroll, C. W., 9, 10, 12
Cauchy, A., 126, 159
Colville, A. R., 168
Courant, R., 6, 8, 10
Curry, H. B., 159

Dantzig, G. B., 6
Davidon, W. C., 161, 194

Edelbaum, T. N., 11
Everett, H. D., 138, 139

Falk, J. E., 15, 126, 130
Farkas, J., 18
Feder, D. P., 157
Fiacco, A. V., 10, 12, 13, 125, 187
Fletcher, R., 135, 157, 161
Flood, M. M., 158
Frisch, K. R., 7
Fromowitz, S., 21

Goldstein, A. A., 12, 159

Hancock, H., 38
Hestenes, M., 38
Hipp, F. C., 128
Householder, A. S., 179
Huard, P., 10, 12, 121
Hurwicz, L., 6, 9

John, F., 38

Karush, W., 38
Kelly, H. J., 11
Kenneth, P., 15
Kripke, B. R., 12
Kuhn, H. W., 6, 20

Leitmann, G., 11
Leon, A., 157, 158

McCormick, G. P., 10, 12, 13, 15, 187
McGill, R., 14, 15
Mangasarian, O. L., 21
Markowitz, H., 186
Martin, A. V., 11
Martin, W. T., 78
Moser, J., 8, 10
Motzkin, T. S., 7
Murray, W., 65
Mylander, W. C., 187

Noble, S. B., 179

Pallu de la Barrière, R., 38
Parisot, G. R., 10
Pennisi, L. L., 38

Pietrzykowski, T., 12, 15
Pomentale, T., 13
Pontryagin, L. S., 15
Powell, M. J. D., 158, 161

Reeves, C. M., 135, 161
Ritter, K., 38
Rosen, J. B., 149
Rosenbrock, H. H., 10
Rubin, H., 8, 10

Spang, H. A., 157, 194
Stong, R. E., 14

Tucker, A. W., 6, 20

Ungar, P., 8, 10
Uzawa, H., 9

Wilde, D. J., 158
Wolfe, P., 9
Wood, C. F., 168
Woodbury, M., 179

Zangwill, W. I., 13, 143, 159
Zoutendijk, G., 149, 193

Subject Index

Acceleration of convergence, *see* Extrapolation
Analyticity, penalty function of convex problem, 110
  trajectory of minima, 78
Arc, *see* Curve
Arrow and Hurwicz method, 128

Bracketing procedure, 192

Carroll's function, *see* Inverse penalty function
Cell problems, 142
Chemical equilibrium problem, 187
Compact perturbation set, 47, 57, 60, 93
Complementary slackness, 19
  strict, 39
Computational aspects, 156
  computer code, 187
  parameter selection, 191
  weights, 114
Computer code implementing SUMT, 187
Concave function, 88
Conjugate direction methods, 158
Constraint qualification, first order, 15, 19, 22
  second order, 25
Continuous version of interior method, 126
  convex problem, 127
  general problem, 126
Convergence criteria, 194
Convex analytic problem, 110
Convexity, exterior penalty function, 102
  functions, 87
  interior penalty function, 94
  programming problem, 89
  sets, 87
Convex programming problem, 3, 86–112
  optimality conditions, 90
  solution by exterior method, 104
  solution by interior method, 97
  solution by Lagrange multiplier dual method, 131
Courant function, 6, 11; *see also* Quadratic penalty function
Criteria for choosing weights, 115
Curve, arc, 19, 20
  derivatives, 20
  tangent, 19, 20

Differences between interior and exterior point algorithms, 63
Differential gradient method, 9
Dual convex problem, differentiable version, 92
  nondifferentiable version, 92
  Wolfe-Huard dual, 121, 130
Duality, dual bounds, by exterior methods, 105
  by interior methods, 98
  dual convex problems, 92
  presetting bounds, 102
Dual Lagrange multiplier algorithm, 131
Dual method, 130
  regularized for convex programming, 136
  for strictly convex programming, 130

Existence of isolated trajectory, 82

207

Subject Index

Exterior point algorithms, 60
  combined with interior methods, 60
  combined with simplicial methods, 145
  convergence for convex problem, 104
  Courant quadratic penalty function, 6, 11
  differences from interior methods, 63
  dual convergence, 105
  general convergence theorem, 57
  intuitive derivation, 53
  T, p, and O functions, 55
  trajectory analysis, 82
  upper bounds, 107
Extrapolation, 72–85
  acceleration of convergence, 188
  in exterior methods, 82
  in interior methods, 72
  in mixed methods, 84
  to next minimum, 189
  to solution, 189
  Vandermonde determinant, 188

Factorable programming, 184
Farkas' lemma, 23
Feasibility, 64, 157, 195
Fibonacci search, 193
Fletcher-Reeves method, 165

Generalized exterior point method, 13
Generalized interior and exterior point method, 65
Generalized inverse, 166
Generalized Lagrange multipliers, 19
Generalized Lagrange multiplier technique (GLMT), 137
  algorithm, 139
  Lambda Theorem, bounds on multipliers, 140
  Main Theorem, 139
Generalized Newton method, 161
Global minimum, 3, 14, 17, 50, 59, 61, 89, 108
Gradient methods, Cauchy steepest descent, 159
  conjugate gradient, 158
  Newton methods, 162
  projected gradient, 149
  reduced gradient, 149
Gradient projection method, 149

$I(x)$, 42; *see also* Interior point algorithms

Implicit function theorem, 35, 80, 82
Interior point algorithms, combined with exterior methods, 59
  combined with simplicial methods, 145
  continuous version, 126
  convergence for convex problem, 94
  differences from exterior methods, 63
  dual convergence, 105
  general convergence theorem, 45
  intuitive derivation, 42
  regularized methods, 50
  trajectory analysis, 72
  U, s, and I functions, 42
Interior point determination, feasibility phase, 195
Inverse penalty function, 13, 41, 79, 158
Isolated set, 46
Isolated trajectory of unconstrained minima, 72

Jacobian matrix, 32, 80, 85, 148

Kuhn-Tucker Necessity Theorem, 20
Kuhn-Tucker Sufficiency Theorem, 90

Lagrange multipliers, 19, 23, 37
Lagrange multiplier technique, 5
Lagrangian, 5, 16, 18
Linear programming problem, 1, 6, 111, 180
Local minimum, 3, 15, 17, 46, 89
Logarithmic penalty function, 40
  presetting bounds, 102
  trajectory analysis, 79
  weighted, 119
Logarithmic potential method, 7
Logarithmic-quadratic penalty function W, 84, 179

Mathematical programming problem, 1
  convex, 89
  discrete, integer, 138
  linear, 1, 6, 111, 180
  nonconvex, 3, 15
  parametric, 34
  quadratic, 3, 16, 186
Method of centers, 10, 12, 121
Methods of feasible directions, 149
Minima, compact set of, 47
  constrained, 46
  global, 3, 14, 17, 50, 59, 61, 89, 108
  local, 3, 15, 17, 46, 89
  unconstrained, 24, 46

Subject Index   209

Minimizing an unconstrained function, 157
　with function values, 158, 175
　with first derivatives, 159, 170
　with second derivatives, 161, 167
　for structured problems, 178
Minimizing on a vector, 191; see also One-dimensional minimization
Minimizing sequence, 47, 57, 60
Minimizing the W function, 178
Mixed interior-exterior point algorithms,
　convergence, 60
　trajectory analysis, 84
Modified Newton method, 167

Necessary conditions, constrained minimization, 24, 25
　convex problem, 91
　first order, 24
　second order, 25
　unconstrained minimization, 28
Negative semidefinite matrix, 89
Neighborhood sufficiency theorem, 91
Newton iteration, 181
Nonconvex programming problems, 3, 15
Nonlinear programming problem, 1; see also Mathematical programming problem

$O(x)$, 55; see also Exterior point algorithms
One-dimensional minimization, bracketing procedure, 192
　Fibonacci search, 193
Optimal search procedure, 193

$p(t)$, 55; see also Exterior point algorithms
Parameters, r, interior methods, 42, 191
　t, exterior methods, 55, 191
　weights, 114
Parametric programming problem, 34
Penalty functions, general, V, 60
　hierarchy of, 68
　inverse, P, 13, 41, 79, 158
　logarithmic, L, 40, 79, 102, 119
　mixed, W, 179
Perturbation of compact convex sets, 93
Perturbation of optimum, 34
P function, see Inverse penalty function
Portfolio selection problem, 186
Positive definite matrix, 32, 88
Positive definite quadratic form, 170
Positive semidefinite matrix, 28, 88

Positive semidefinite quadratic form, 116
Primal convex programming problem, 92
Primal-dual bounds, 92
Problems, A, mixed equality-inequality, 1
　B, inequalities, 4
　C, convex primal, 89
　D, convex dual, 92
　E, equalities, 11
　F, mixed equality-inequality, 14
　NL, mixed nonlinear-linear, 146
　P, parametric, 34
　β, dual, 130
　$\beta(\epsilon^k)$, perturbed dual, 136
　$C(\epsilon^k)$, perturbed primal, 136
Projected gradient method, combined with penalty method, 149
　projection matrix, 149
Pseudoinverse, generalized inverse, 21

Q-function type algorithms, 121; see also Method of centers
Quadratic penalty function, 8
　Courant function, 8, 11, 14, 64, 82
Quadratic programming problem, 3, 186
　arising in penalty-weight selection, 116

Rank annihilation method of matrix inversion, 179
Reduced gradient method, 149
Regularized interior point methods, 50

$s(r)$, 42; see also Interior point algorithms
Saddle-point problem, 6
Search techniques, 158
Sensitivity analysis, 34
Separable programming problems, 3, 183
Sequential unconstrained minimization techniques, 3
　exterior algorithms, 53
　interior algorithms, 39
Shadow prices, 37, 141; see also Lagrange multipliers
Simplicial algorithms, combined with penalty methods, 145
Steepest ascent, 141
Steepest descent, 126, 157, 159
Strict complementarity slackness, 81
Strict convex problem, 131
Strictly convex function, 88
Structured problems, 178
　factorable, 184

linear, 180
separable, 183
Sufficiency conditions, constrained minimization, 30
  convex programming, 90, 91
  neighborhood, 33
  second order, 30
  unconstrained minimization, 31
  use in trajectory analysis, 80, 84
SUMT, 187

$T(x,t)$, 55; see also Exterior point algorithms
Tangent, 20
Trajectory analysis, logarithmic penalty algorithm, 79
  mixed penalty algorithm, 84
  simplicial-penalty algorithm, 147
Trajectory of minima, analytic, 78
  differentiable, 77
  isolated, 73

$U(x,r)$, 42; see also Interior point algorithms
Unconstrained minimization algorithms, comparison of methods, 162
  computational aspects, 156
  conjugate gradient, 160
  Fletcher-Reeves, 161
  generalized Newton, 161
  GLMT, 137
  modified Newton, 167
  search, 158
  steepest descent, 159
  for structured problems, 178
Unconstrained minimization, solution of constrained problem, 143
Unconstrained minimum, 2, 24
  global, 86
  local, 28, 30
  of penalty functions, 46
Unconstrained-simplicial algorithms, 145
Upper bounds for exterior algorithms, 107

$V(x,r,t)$, 60; see also Mixed interior-exterior point algorithms
Vandermonde determinant, 188
Variable metric method, 161
  Davidon, 161
  revised, 170

Weights in penalty function, 114
  computational advantages, 115
  criteria for selection, 116
W function, 179